SACRED GEOMETRY
OF THE
EARTH

"The intriguing theories of Catherine Young and Mark Vidler concerning the coincidence of prominent landscape features shown in their juxtapositions to express mathematical ratios in terms of nautical miles should not be dismissed."

JOHN NEAL, AUTHOR OF *MEASURING THE MEGALITHS* AND *THE STRUCTURE OF METROLOGY*

"An astonishing piece of work and utterly compelling."

ALLAN BROWN, ILLUSTRATOR AND CONTRIBUTOR TO *HOW THE WORLD WAS MADE*

SACRED GEOMETRY
OF THE
EARTH

THE ANCIENT MATRIX
OF MONUMENTS
AND MOUNTAINS

MARK VIDLER
AND CATHERINE YOUNG

Inner Traditions
Rochester, Vermont • Toronto, Canada

Inner Traditions
One Park Street
Rochester, Vermont 05767
www.InnerTraditions.com

SUSTAINABLE FORESTRY INITIATIVE   Certified Sourcing
www.sfiprogram.org
SFI-00854

Text stock is SFI certified

Library of Congress Cataloging-in-Publication Data
Vidler, Mark, author.
  Sacred geometry of the Earth : the ancient matrix of monuments and mountains
/ Mark Vidler and Catherine Young.
    pages cm
  Includes bibliographical references and index.
  ISBN 978-1-62055-468-5 (pbk.) — ISBN 978-1-62055-469-2 (ebook)
  1. Megalithic monuments. 2. Neolithic period. 3. Archaeoastronomy. I. Young,
Catherine, 1958– author. II. Title.
  GN790.V54 2016
  930.1'4—dc23
                                                                      2015013859

Printed and bound in the United States by Lake Book Manufacturing, Inc.
The text stock is SFI certified. The Sustainable Forestry Initiative® program
promotes sustainable forest management.

10 9 8 7 6 5 4 3 2 1

Text design by Priscilla Baker and layout by Debbie Glogover
This book was typeset in Garamond Premier Pro with Gill Sans as display font
Artwork by Mark Vidler and Debbie Glogover

To send correspondence to the authors of this book, mail a first-class letter to the
authors c/o Inner Traditions • Bear & Company, One Park Street, Rochester, VT
05767, and we will forward the communication, or contact the authors directly at
mpv123@gmail.com.

For Isla and Katy
and
Laura and Emma and Will

CONTENTS

ACKNOWLEDGMENTS

Special thanks to Rand Flem-Ath for his insight and enthusiasm. We would also like to thank Rose Flem-Ath, Lizzie Hutchins, Oliver Bentham, Allan Brown, Paul Cooper, Sheila Knight, Tessa Rawcliffe, Laurence Glazier, Andrew Pickersgill, Emma Young, Gary Heidt, and Jon Graham. We are indebted to cartographers through the ages and all the people who have contributed to the creation of Google Earth.

Keys to Earth's Sacred Geometry

My adventure exploring the "ancient matrix of monuments and mountains" described in this book began in June 2010, when, no longer able to put off searching the attic for something I've long since forgotten I wanted, I stumbled over that most enticing of treasures for a librarian— a box of books. Brushing off the dust and squinting through a tangle of spiderwebs, I saw the blue and silver cover of *The Star Mirror* by Mark Vidler. It had been on my must-do reading list ever since an editor at the London offices of HarperCollins had given it to me when Colin Wilson and I were making presentations for *The Atlantis Blueprint*. I brought Mark's book home to Canada, but before I had a chance to delve into it, my own book deal was signed, and writing, researching, and editing pushed the luxury of reading for pleasure to the bottom of the list.

After I finally pulled *The Star Mirror* into the light of day, I was astonished at what I found and regretted that I hadn't read it sooner. Within its pages Mark had judiciously challenged common assumptions about our planet's place in the universe. We soon began a lively and fascinating e-mail conversation. His work was the most exciting I'd found since I began corresponding with Charles Hapgood.

After repeated investigation Mark had discovered that 3,000 kilometers was the length of a straight line drawn between many notable sites (both natural and manmade) around the Earth. For instance, 3,000 kilometers separate Mexico's Pyramid of the Sun from the northern peninsula of South America. He had also discerned the critical importance of the cardinal points (extreme north, south, east, or west) of the continents.

Fascinated by the material I dug deeper into Mark's 3,000 kilometer figure. I knew that the kilometer—a measurement created in relatively recent historic times in Napoleonic France—was unlikely to have been used by the architects of ancient Egypt and Mexico whose sites Mark was exploring.

Living on an island where we're dependent on ferry service for everything from cameras to coffeepots and where boats are almost as common as cars, I was curious to translate 3,000 kilometers into nautical miles. (I had made extensive use of the nautical mile in countless calculations for the writing of *The Atlantis Blueprint*.) Although the nautical mile is thought to be a modern invention, determined in 1929, I had concluded as early as 1976 that our planet's true dimensions were known in very ancient times. And since the nautical mile is a precise subdivision of the actual dimensions of the Earth,[1] its use in ancient times would indicate a greater knowledge of geography than is normally granted to any ancient civilization. When I converted Mark's 3,000 kilometers into nautical miles, I was amazed at the results and immediately wrote to him, pointing out how close 3,000 km is to 1,618 nautical miles (or the universal constant phi multiplied by 1,000).

The realization that 1,618 nautical miles is intimately related to the universal geometric constant phi was compelling evidence that the common distances linking sacred sites, mountains, and cardinal points that Mark had found all over the globe were far beyond coincidence.

I wrote to Mark: "I can't help but wonder if there isn't a π equivalent. It might be interesting to draw 3,142 nautical mile circles around the various mountains and sacred sites and see what they connect with." It wasn't long before Mark demonstrated conclusively that pi

(3.142) multiplied by 1,000 and measured in nautical miles yielded a second key for unlocking the sacred geometry of the Earth. For example, it is 3,142 nautical miles from Mexico's Pyramid of the Sun to the most northerly peninsula of North America's landmass.

Sacred Geometry of the Earth reveals that pi and phi numbers (3,142 and 1,618) are embedded into the body of the Earth like an ancient scar. These iconic numbers trace nautical miles that link some of the most mysterious places on the planet. They set compass bearings between sites with incredible accuracy and relentless consistency. The very existence of such sophisticated knowledge signifies that our concepts of the past are more than flawed, they are wrong. The time is overdue for the exciting new understanding that can be found on every page of Mark and Catherine's book.

RAND FLEM-ATH

RAND FLEM-ATH is a librarian and coauthor of *The Atlantis Blueprint*. He has appeared on the History Channel, the Discovery Channel, NBC, CBC, and the BBC as well as on numerous radio shows. He lives on Vancouver Island in British Columbia.

Le Menec stone rows at Carnac.
Postcard photo by Jos Le Doaré.

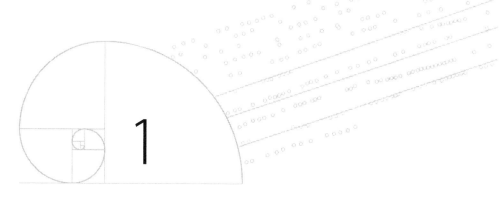

SIGNPOSTS IN THE
LANDSCAPE

LINES FROM THE PAST

The stone rows shown in figure 1.1 are called Le Menec; they are part of the largest Neolithic monument in Europe at Carnac, in the Brittany region of France. A series of lines like these spans over three kilometers, yet these alignments have remained an enduring mystery: why would anyone do this?

The people who labored to create Le Menec left no written record to explain their motivation; today we can only guess at what they were trying to achieve. Our guesses have been imaginative; the stones form meeting places, temples, or astronomical observatories. They are fertility symbols, geometric diagrams, sources of magic, and so on.

But at Carnac there is a possible clue. Although the dilapidated rows at the beginning of Le Menec do not follow precisely straight lines, they do indicate a general direction. Are they a signpost? Do these lines indicate a direction to follow?

The straight lines A, B, and C in figure 1.2 diverge slightly as they are extended eastward. If continued they would reach the following destinations:

Figure 1.1. The beginning of Le Menec at Carnac, in the Brittany region of France; the standing stones were probably erected between 4000 and 3000 BCE. Photo by OliBac.

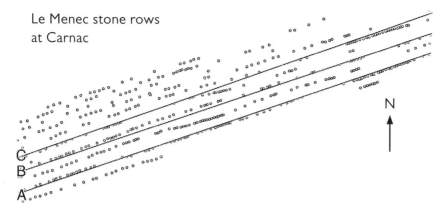

Le Menec stone rows
at Carnac

N

C
B
A

Figure 1.2. The stone rows at Le Menec; each standing stone is plotted as a circle. The image covers the first five hundred meters of the alignments. Many of the stones have been re-erected. The linear clusters to the east have certainly been re-erected.

A. The summit of Mount Everest.
B. The summit of K2.
C. The summit of Mount Kanchenjunga.

The Neolithic lines are directed toward the three highest mountain summits in the world. It's going to be a long walk, and to follow these lines we do not need a map but a globe.

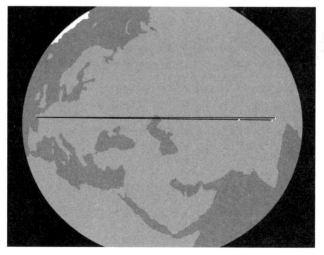

Figure 1.3. The lines from Le Menec extend to Mount Everest, K2, and Mount Kanchenjunga.

COINCIDENCE?

To consider that these alignments were intended would imply that someone in prehistoric Europe knew where the world's highest mountains are located. Such an idea would be regarded as highly improbable, if not entirely impossible, by a majority of historians today. But it is equally improbable that randomly located lines on the Earth should so readily extend to its three highest summits by chance.

MAN-MADE MOUNTAINS

Overlooking the stone rows at Le Menec is one of Europe's largest Neolithic mounds, an enormous man-made hill called St. Michael's Tumulus. If this mound is interpreted as a symbolic mountain (^), then the stone rows of Le Menec serve the purpose of joining four mountain symbols together: ^ St. Michael's Tumulus, ^ Mount Everest, ^ K2, and ^ Mount Kanchenjunga; this can be represented by ^-^-^-^ on the printed page.

If this alignment of high points was intended, it not only demonstrates a sophisticated understanding of world geography in the distant past, but it could also suggest a method of recording information by joining symbols on lines.

As a stand-alone example of Neolithic ability, the alignments of Le Menec with the Earth's highest mountains could be understood as an odd coincidence and any suggestion of a language of symbols easily dismissed. But Carnac is no stand-alone example. Alignments like these are replicated from one Neolithic monument to the next. There is clear evidence of a common understanding among the ancient monument builders.

At the beginning of Le Menec, there is a large stone circle big enough to enclose several houses. This circle defines a geometric point in the landscape, a point from which the straight lines are "drawn" in stone. Why would anyone in prehistory exert so much effort to create geometry in the landscape?

ANCIENT CONSENSUS

It cannot be known what the Carnac builders were thinking because they left no conventional written record, but as literacy gradually developed and ideas were written down, numerous myths from across the ancient world reveal a notable consensus of belief. In Europe, Africa, and India, the understanding is very similar to that of the Native Americans and the Australian Aboriginals. In varied and various ways, people from all around the ancient world expressed the belief that the Earth was something "created." The same understanding permeates the Bible, the Qur'an, and other religious texts. Creation myths describe omnipotent forces bringing order out of chaos, of raising the Earth from the primordial soup, or creating the planet in a manner too supernatural to envisage. Accounts vary considerably, but at their core they reveal a widespread ancient understanding that the Earth is not an "accidental agglomeration of dust particles," as we have it today. Rather the opposite. For example, the ancient view described in the *Hermetica* is one of a "*fashioned*" object that is "*structured*" and "*ordered.*"*[1] Words like *created, formed,* and *made* are also common among the creation myths. So where did the idea of a "designed world" come from, and could the lines drawn across the landscape at Carnac have anything to do with it?

Despite the fact that many of these accounts express similar underlying views, each is frustratingly nonspecific and short on detail. They offer no direct evidence to support their claim that the planet was created in some mysterious way.† In the *Hermetica* Hermes Trismegistus reputedly teaches the philosophy of Hermes. He says the Earth is "fashioned" and the proof of this can be found "everywhere," yet no direction to evidence supporting this belief is given.

---

*For instance, "And when the Creator had made the ordered universe, he willed to set in order the earth also." The word *ordered* is used repeatedly in Hermes' teaching with regard to the Earth, the universe, and everything in it.

†There are exceptions, which include the Australian Aboriginal creation myths; see chapter 15, "Pi and the Songlines."

The Bible says that "the foundations of the earth" were divinely appointed, but no plan or blueprint is provided.[2] In the Qur'an it says, "He it is who hath outstretched the earth, and placed therein firmly set mountains."[3] Here the mountains are divinely located to stabilize, or peg down, the Earth. In the earliest Indian texts, it is Varuna, "the Lord Immortal," who "has meted the earth out," that is, apportioned the Earth.[4]

These fundamental beliefs emerged with recorded history. Could it have been the people of the Stone Age who initially gave rise to this understanding? If the Neolithic monument builders also believed in a designed and ordered Earth, then the locations for their lines, circles, and man-made mountains should surely be integrated into the perceived natural design, rather than being carelessly scattered over it. Could the motivation to build Le Menec be understood as part of a measured design?

SOMETHING MEASURED

A number of ancient accounts suggest the Earth was not only created, but was also carefully measured. The Apache describe different colored cords being stretched to the north, south, east, and west of the world, and this vision is echoed in the *Book of Enoch*. This gnostic text describes a mysterious survey of the Earth, albeit in flowery language.

> And I saw in those days how long cords were given to the Angels, and they took themselves wings and flew, and went towards the north. And I asked an Angel, saying unto him: "Why have these taken cords and gone off?" And he said to me: "They have gone to measure."[5]

From as far afield as the Quiche kingdom in the western highlands of Guatemala, they apparently "examined the four corners, the four points of the arch of the sky, and the round face of the earth."[6]

The Mexican god Quetzalcoatl was said to have been "the great one who first measured the earth."*[7]

There is a similar reference in the Mayan holy book, *The Chilam Balam of Chumayel*. "'Who has passed here? Here are footprints. Measure it off with your foot.' So spoke the mistress of the world. Then he measured the footstep of our Lord, God the Father. This was the reason it was called counting off the whole earth."[8]

Julius Caesar recounts that among the Druids he met in Britain (55–54 BCE), no writing was permitted in the instruction of sacred subjects. Youths were taught about "the dimensions of the world and of countries." Everything had to be learned by heart and passed down by word of mouth, generation to generation. Writing was considered to impede memory in discussions relating to "the size of the universe and of the earth."[9]

The basic tool needed to measure the size of something is a straight line, and a cord stretched between two posts is a method still employed today. But Caesar's account says much more than this; he tells us specifically that the dimensions of countries were known and that this knowledge of the Earth's topography had been passed down from generation to generation. The information was clearly considered valuable, but he offers no hint that the Druids could make these measurements themselves.

THE ACCIDENTAL EARTH

Accounts of the Earth being designed and descriptions of people measuring it in antiquity were gradually dismissed or forgotten as history began moving into the modern age. Sir Isaac Newton (1642–1727) believed the dimensions of the planet were known by the Egyptian pyramid builders, but as time passed the belief in "ancient wisdom" declined and the poetic vision of a designed Earth suspended in an ordered universe was gradually superseded by a scientific one. As with

---

*Bernardino de Sahagun (sixteenth century) reports that Quetzalcoatl "brought the skills and sciences necessary to create civilized life, introducing the secrets of masonry and architecture, heralded as the father of mathematics, metallurgy, and astronomy, the great one who first measured the earth."

any shock acceptance took time and the change of perspective was gradual. Ushered in by Copernicus in 1543, then laid low by Newton's *Principia* in 1687, perhaps the final death knell for the old belief was sounded in Charles Lyell's *Principles of Geology*, published in 1830. Following Lyell's bestseller there was little room left for thoughts about a "created earth" when his book showed that it had evidently formed fitfully over tens of millions of years. The belief in the "divinely made man" and the "divinely made earth" were both gradually worn down by science, and the once hallowed vision of a "grand design" was ultimately undone by a few crates of fossils.

Today the Earth's origins are seen as a random accretion of dust, described here by Professor Ian Stewart.

> The number of planets presumably depended on the amount of matter in the dust cloud, how it was distributed, and how fast and in what direction it was moving . . . this could have given us eight planets, or even eleven; the number is accidental.[10]

The modern view of an "accidental" Earth could scarcely be more distant from the ancient vision of an ordered one.

BAD TIMING

The odd thing about this change of perspective was its timing. When Lyell's book was published in the early nineteenth century, many parts of the world remained unexplored and the limits of the continents were, in many cases, still uncharted. The North American coast was unresolved, chunks of Africa were unknown, and the location of the world's highest mountain was still a mystery. No accurate overview of the planet was available at the time, and consequently there was little possibility of discovering whether the ancient idea of an ordered Earth could be verified.

Despite the rapid ascent of scientific understanding, it has only recently become possible to look at the entire planet on a digital spherical map, Google Earth. This high-tech tool allows us to view the Earth from a number of perspectives; many people check out their house and neighborhood in "street view," but the satellite images show the entire topography of the Earth in unprecedented detail, including the ancient monuments located on it. At worst, the view this affords can be described as interesting; at best, astounding. The "ruler" provided on Google Earth draws and measures straight lines from point to point on the surface. Clearly visible on these satellite images is the great man-made mound at Carnac, with the chapel of St. Michael on its summit. This overlooks the Le Menec stone rows and the circle where the rows begin. The ruler can be used to draw the straight line from Carnac to Mount Everest. Google Earth also provides a detailed view of Europe's largest Neolithic mound, Silbury Hill at Avebury, England. So if the alignments at Carnac were intended, does Silbury Hill also align with high points?

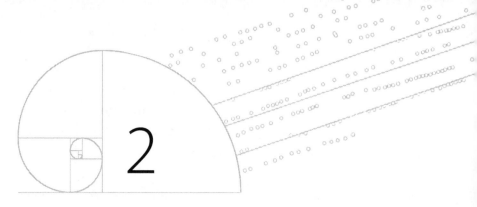

2

PATTERNS IN THE HILLS

A CALM CROSSING

It is a calm crossing, the low tide washing the hide hull gently onto the moonlit mudflats. They are among the first to make this journey, never to return. Their mission is to hack trenches in the hills. These boat people will carve a new landscape. They come in their hundreds, moving westward overland and drifting to the north, deflowering the virgin hills in their wake. Great hunks of land are banked up, ditches are hacked into rock; millions of tons of earth are shifted. Countless hours of labor will change the skyline forever. But is there measure in this madness? A mysterious calculator moves from hill to hill. Some hills are cut at the summit, others just below, others on the shoulder. The year is 3800 BCE, and the work has only just begun.[1]

Two thousand years pass by, and the landscapes of Britain and Ireland have changed forever. Hundreds of hills and valleys are now grooved or sliced with meandering banks and ditches, often completing a circuit; some are circular. The Neolithic vision remains deeply etched. Archaeologist Aubrey Burl suggests that during this period as many as four thousand stone circles may have been constructed; certainly one thousand were, along with numerous stone rows, countless standing stones, dolmens, cairns, and barrows. And among these monuments a

Figure 2.1. Silbury Hill at Avebury in Wiltshire, England.
Photo by Chris Talbot.

handful of truly enormous earth mounds were created. The largest of these is Silbury Hill, located at Britain's largest Neolithic site, Avebury in Wiltshire.

A VIEW OF AVEBURY

Today Avebury offers a quintessential view of England: sheep nibble unconcernedly around the ancient gray stones; village houses, no two the same; honey-hued barns and deeply rutted farm tracks wind up to the North Wiltshire Downs; rounded green hills, some crowned with a copse. But this tranquil landscape once bore witness to a Neolithic building program of astonishing proportions. The outer stone circle at Avebury is the largest in Europe, and the great surrounding henge (a near circular ditch) is the deepest in Europe. How was this location chosen, and why was it so important?

The complex of monuments at Avebury once included not only the main henge and a large perimeter stone circle, but also two smaller central circles contained within it. From there two snaking avenues of stones stretched over the neighboring slopes, one ending in a small outlying circle known as the Sanctuary. There were once six hundred standing stones at Avebury, but between the fourteenth and nineteenth centuries many were systematically buried or broken, and fewer than one hundred remain today. The Avebury complex includes a number of large long barrows and several cairns, but all this is dwarfed by the massive, monumental mound of Silbury Hill, about 1,200 meters south of the main henge.

This majestic man-made hill is described in the local Alexander Keiller Museum as "the highest artificial pre-historic earth mound in the world." Silbury Hill is more than 4,300 years old. It stands 30 meters high and is 160 meters wide at the base. It consists of approximately five hundred thousand tons of material,[2] yet its true purpose and significance remain unknown.

The motivation for building Silbury Hill remains a mystery, but if Neolithic people held the belief that high points on the Earth are especially ordered, this might explain the commitment of thousands of hours of labor to build a hill where none previously stood. It could be an homage of some description, a replication or reflection of the natural pattern of hills. But if so, how would they decide where to put it?

STRAIGHT TRACKS

Alfred Watkins was an eminently practical man who paid careful attention to detail. In his teenage years his dedication to photography, in the pursuit of which he would spend hours developing wet–plate glass negatives in a little tent, led him to the discovery that successful photographs were a matter of correct exposure and development times. Herbert Ponting, who accompanied Captain Robert Falcon Scott's Terra Nova expedition to the Antarctic, took with him a small device of Watkins's invention that would calculate these key times. Many of the iconic images of Scott's expedition are a direct result of Watkins's Bee Meter.

When he grew up Watkins became an "outrider" for the family businesses in Herefordshire, England. On his travels around the county, he made what to him was a startling discovery. He envisaged a series of straight lines that appeared to link various landmarks and looked rather like an ancient network of paths. Watkins believed that these alignments could not be due to mere chance. He thought prehistoric man had deliberately made tracks to provide a network of straight paths over many miles of the English countryside. He described these paths as "ley lines" in his book *The Old Straight Track,* published in 1925.

The *Collins English Dictionary* defines a *ley line* as "a line joining two prominent points in the landscape, thought to be the line of a prehistoric track."

As his work progressed Watkins concluded that ley lines often intersected or terminated on what are known as "beacon hill" summits, generally the highest summit or most prominent point in a region. Subsequent research has concluded that the ancient landmarks adopted by Watkins are so numerous that it is no surprise to find lines that join them together. So Watkins's understanding that these lines were directed at regional high points in the landscape is now often overlooked.

INVISIBLE LINES

In the decades since Watkins coined the term *ley line,* the true meaning has become so badly misunderstood that although you can stick a spade in them, the very existence of these lines is often denied. In his book *Archaeological Theory: An Introduction,* Matthew Johnson says simply, "Ley lines do not exist."[3] Why the denial?

Part of the problem may be attributed to the association of ley lines with Earth energies, divining, and ufology. In the 1960s the ley line was almost completely lost in a froth of new age ideas, now reflected in the *Oxford Dictionary* definition of a ley line, which makes no mention of their alignment to prominent points: "A supposed straight line connecting three or more prehistoric or ancient sites, sometimes regarded as the line of a former track and associated by some with lines of energy and other paranormal phenomena."

Wow! Think flower-bedecked hippies, possibly smoking interesting organic matter and offering free love, or Druids and white witches bonding with the Earth forces of Mother Nature. Chuck in a few doubtlessly dodgy aliens zipping in and out of orbit, and what's not to run away from? Academics can hardly be blamed for divorcing themselves from any association with ley lines, but there is a more fundamental reason for their censure.

The implication that Neolithic people were capable of creating long-distance alignments is anathema to orthodoxy. Archaeologist Aubrey Burl is clear: Neolithic people "were not geodetic surveyors,"* and he believes, "True North or South were abstractions and matters of indifference, even incomprehension, to them."[4] If such staunchly conservative ideas still dominate the discipline of archaeology, they will discourage any serious investigation of long-distance measurements.

LONG MEASURES

The complex of monuments at Avebury can be seen on a single Ordnance Survey map, and it is here that Watkins's promised alignment with high points in the landscape immediately comes to life. When lines are drawn from one natural hill summit to the next, the ancient vision of an ordered landscape emerges around Silbury Hill. The evidence is on the topographical map.†

In a ten-kilometer radius of Silbury Hill, the highest summit to the north of the monument is Hackpen Hill at 273 meters. In the same radius to the south are the two highest summits on the North Wiltshire Downs: Tan Hill is 294 meters high, and its neighbor Milk Hill is 295 meters. A single straight line drawn on the map clearly shows that

---

*A geodetic survey takes account of the Earth's curvature.
† Ordnance Survey Explorer map no. 157, scale 1:25,000

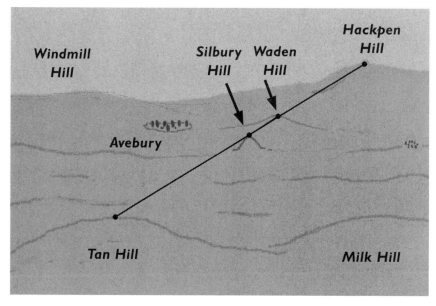

Figure 2.2. The illustration shows that Europe's largest Neolithic mound is on the line joining two dominant summits in the area, and the nearest hilltop adjacent to the monument is also on this line.

Silbury Hill is aligned directly between the summit of Hackpen Hill to the north and the summit of Tan Hill to the south.* When the modern trig point markers on both summits are joined with a line, it passes over Silbury Hill. The line also passes over the summit of Silbury Hill's nearest neighbor, Waden Hill (see figure 2.2).

The location of Silbury Hill may have been chosen to reflect, duplicate, or perhaps simply add to the alignment of three natural high points. If this was intended by the builders, it draws attention to a simple geometric arrangement of three natural summits on a straight line, that is, the apparent ordering of hills in the landscape.

Might this alignment suggest a purpose for the monument's construction? A possible answer is offered by looking at the broader distribution of the highest natural summits in relation to Silbury Hill.

---

*Both summits are relatively flat.

THE BROADER PICTURE

What cannot be seen directly from Avebury is the larger geographical picture. The helpful information board in the parking area indicates that "You Are Here," but only links the visitor with the local topography, not the national. If the alignment of high points found locally at Avebury was intentional, how far across the country might these hilltop alignments extend?

To the east, about thirty kilometers from Silbury Hill, is the highest summit in central southern England, Walbury Hill, with a height of 297 meters. This hilltop is almost flat, like a small plateau, with the land descending sharply all around its perimeter. Further to the east there is only one summit of comparable height, Leith Hill, at 294 meters. About eighty kilometers from Walbury Hill, Leith Hill overlooks all of southeast England.

Although these two regional high points are about eighty kilometers apart, a line drawn from the summit of Silbury Hill to the summit of Leith Hill passes directly over the flat summit of Walbury Hill.

So Silbury Hill, the largest Neolithic hill in Europe, is aligned with the highest summits in central southern England and in southeast England, an area of about 25,000 square kilometers.

The highest summit in all of southern England is High Willhays on Dartmoor in southwest England. From there a line drawn to Avebury continues to the highest point in the Chiltern Hills, Haddington Hill.

To give an idea of the prominence of all four of these summits, figure 2.3 shows southern England divided into four regions; these four hills are the highest in each of the outlined regions.

Each one of these four dominant summits is a "regional high point"; when lines are drawn between these points, Avebury is in the crosshairs (figure 2.4).

Locating the precise summit of a hill can be tricky. Nevertheless, it is true to say that these two lines joining the four regional high points across the country intersect at Avebury. Is it simply chance that geometric points located on these four regional high points can be

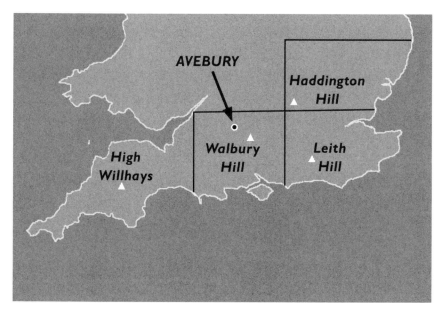

Figure 2.3. Four areas of southern England are outlined, and the highest summit in each area is named and marked with a white triangle.

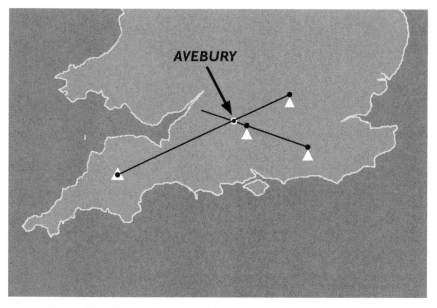

Figure 2.4. The four regional high points are joined in pairs. The lines intersect at Avebury.

joined together to create two isosceles triangles? Was Avebury located in alignment with these hills because of the geometry that can be created between them?

EXTREME POINTS

With his Bee Meter and other inventions, Alfred Watkins was clearly an innovative individual, but he could not extend lines right across the country as we can today using Google Earth. If the lines intersecting at Avebury are extended to the coast, the result is a surprise. *Both ends of both lines* reach the sea at notable coastal points (marked A, B, C, and D in figure 2.5).

The line from A to B shown in figure 2.5 begins at England's most westerly extreme, the Land's End Peninsula, and it ends at England's most easterly extreme, Ness Point in East Anglia. This line passes directly over southern England's highest point, High Willhays, and also over the

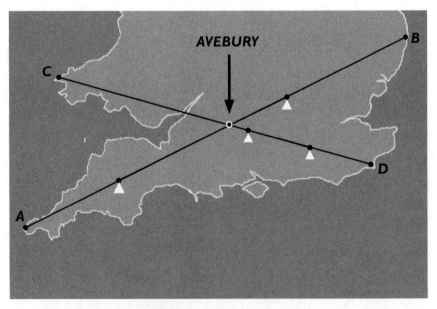

Figure 2.5. The alignment of Avebury, in the crosshairs, with straight lines joining regional high points (internal dots). The straight lines extend to four coastal "points" (coastal dots).

Avebury complex. If this line is "read" as a line of symbols, ^-^-^, then two further topographical extreme points have now been added to it.

The line from A to B now reads:

Cardinal coastal extreme point	^
Regional high point	^
Avebury complex	^
Regional high point	^
Cardinal coastal extreme point	^

A line from Ness Point passing over the Sanctuary at Avebury extends to the base of the Land's End Peninsula; and a line from Ness Point over High Willhays summit extends to the tip of the Land's End Peninsula.*

The line from C to D is very similar. It begins on the landmass of St. David's Head, at the western extreme of Wales, and then passes over Silbury Hill. It continues over the two regional high points of Walbury Hill and Leith Hill before terminating on the flank of the Dungeness Peninsula in the south. The Dungeness Peninsula is a notable chevron-shaped landmass on the south coast "pointing" several kilometers into the sea (marked D in figure 2.5). Like flags dotted around a golf course, chevron-shaped peninsulas appear to be regularly targeted by Neolithic alignments. The peninsula at Dungeness is created by approximately twenty-five square kilometers of shingle, thought to be one of the largest shingle beds in the world; a great place to fly a kite.

Buffeted relentlessly by wind and water, coastal extreme points may be subject to more rapid erosion than hilltops. We are not suggesting that coastal extremes like Dungeness were necessarily in their current location five thousand years ago, but we will see that geometric points located on these coastal extremes create surprisingly well-ordered

---

*The Avebury Sanctuary is on the line joining Silbury Hill, Walbury Hill, and Leith Hill. The line from the Land's End Peninsula to Ness Point also passes over the Sanctuary, but only when the line is drawn from the base of the Land's End Peninsula. Base-to-tip measurements are discussed in some detail in the text and the appendices.

geometry in the landscape today. It is mysterious that a number of monuments identify exactly the same coastal extreme points through geometry and alignment. For example, the geometric point located in the east, at Ness Point, aligns with Avebury and the highest point in southern England, and yet exactly the same geometric point is found at one corner of a right triangle, with the other two corners on the tip of the Dungeness Peninsula and on the summit of Silbury Hill. So the high points aligned with Avebury create isosceles triangles in the landscape, and two coastal extremes at the end of the lines create a right triangle with Silbury Hill. During the thousands of years since Avebury's creation, sheep have grazed placidly among the great standing stones. How many of them, in their munching meanderings, have paused at the intersection of these lines, blissfully unaware of the unusual nature of their location?

But Neolithic people were not sheep.

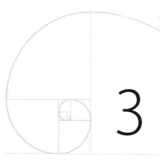

3

MEASURED MONUMENTS

PROFESSOR THOM

In 1974 a visitor to Avebury might have caught sight of the wiry figure of Professor Alexander Thom taking measurements with a theodolite. Or the professor might have been "shooting the sun" to synchronize with the pips of Greenwich Mean Time on his portable radio. Having established true north ("plus or minus 5 arc seconds"), he would have then run what is known as a "closed seven-sided traverse around the ring," and ultimately measured all the angles required to pinpoint the stones in the Avebury circle with great accuracy. The circle at Avebury is about 450 meters across. Thom carefully measured and remeasured distances with steel tapes carried in a backpack by his son and assistant, Archie.

Thom, Emeritus Professor of Engineering at Oxford University, was making the most accurate survey of the Avebury stones ever undertaken. With his pipe tucked between his teeth and his boots shining with the morning dew, he doggedly proceeded with the task of plotting each stone and each of the concrete markers erected where stones had once stood, ninety-four points in total. His pocket notebook was gradually filled with coordinates and measurements that were taken to the local Alexander Keiller Museum, where a large table had been prepared for him. There, spread out over the table, the great scale plan of the ancient stone circle would reach fruition. The result was a masterpiece, but it was not of Thom's making.

The stones at Avebury do not form a circle. They follow a series of joined arcs* with different diameters. Thom located the various centers of the arcs. When three of these points were joined together with straight lines, they created a 3-4-5 Pythagorean right triangle.

The unit of measurement used to create this triangle was already known to Thom; he had found it all over the British Isles, among the hundreds of sites he surveyed. He called this unit a megalithic yard, which is 0.8291 meters. Circumference measurements yielded units of 2.5 megalithic yards, and Thom named this unit a megalithic rod.

THOM'S CIRCLES

Few people were better qualified to investigate this geometry, yet even Thom himself could not understand how the Neolithic builders had managed to retain their regime of whole number measurements at every turn. The series of seven arcs created a closed space, but the length of each arc was "in every case close to an integral number of [Megalithic] Rods." Writing in *Megalithic Remains in Britain and Brittany,* Thom said, "It will be understood that to invent a design dependent on a geometrical construction with this peculiarity presented a very difficult problem."[1]

By all accounts Thom's idea of a "very difficult problem" is likely to be an insurmountable one to many laymen and scholars alike. Nevertheless, there is clear evidence in Thom's plan of Avebury that this Neolithic monument should be considered as the work of extraordinarily sophisticated minds. In an interview for the BBC, Thom summarized his view by saying that the Neolithic builders were "as far as brain power is concerned, my superiors."[2] Yet they could apparently neither read nor write.

Arc: in this case, a portion of a circle's circumference.

Thom was a highly respected engineer, a polymath who taught both engineering and surveying and wrote numerous published papers on field-working methods. Not one to express lightly the brain power evident in the Neolithic Period, he came to this conclusion following arduous calculations based on data collected from hundreds of monuments over a period of forty years.

NO MEAN TASK

Having surveyed Britain's third largest stone circle, the Ring of Brodgar in Orkney, Thom concluded the builders had used pi at a value of 3.14 and had aligned the stones to the North and South Poles with an accuracy of eight arc minutes, which was "no mean task," as he wryly observed. Thom believed accuracy in measurement was akin to a Neolithic signature because he found it at so many of the sites he surveyed.

Thom's vision erases the common image of the Neolithic savage and replaces it with an intellectual giant whose monuments are manifestations of a "megalithic science," the complexity of which is revealed in the pages of calculus that pepper his work. He and his son wrote three books describing his research, but none was suited to the prevailing archaeological viewpoint. Nor are they to this day—because they describe an early culture with profoundly advanced abilities in surveying, mathematics, geometry, and astronomy.

The idea of a hidden Neolithic intellect revealed through accurate measurements stands as a bold alternative to the orthodox understanding. Archaeological excavations have already established that Neolithic people were essentially bone-chewing peasants with little taste in earthenware and a tendency to violence. Thom's surveys provide a statistically supported contradiction to this view, but since his death in 1985 critics have found a handful of errors in his work. Homing in on these alone, the majority of archaeologists have dismissed the rest of his work; it is now almost completely bypassed by mainstream academics.

However, a recent archaeological discovery places a new perspective on Thom's findings: Silbury Hill has a partner: the Marlborough Mound (figure 3.1).*

SILBURY HILL AND MARLBOROUGH MOUND

Until recently Silbury Hill was often assumed to be the only substantial Neolithic mound in the Avebury area, but this viewpoint has now changed. About eight kilometers to the east of Silbury Hill is the Marlborough Mound, the second largest earth mound in England. The Marl Burgh, as it is sometimes called, was once assumed to have been constructed by the Normans, some 3,500 years after Silbury Hill.

In 2011 charcoal fragments were taken from the nineteen-meter-high spiral mound, and radiocarbon dates indicate that it is about 4,500 years old. The dig leader, Jim Leary, described the results as "an astonishing discovery." The carbon dating revealed that the Marlborough Mound and Silbury Hill could have been constructed as part of a single project. So are these two monuments like two posts in a survey, two markers ordered in accordance with some form of order between hills on the Earth?

USING THOM'S MEASUREMENTS

The distance between Silbury Hill and Marlborough Mound was calculated using a Garmin Global Positioning System (GPS). A straight line of 10,000 megalithic yards can be measured between them, from the summit of Marlborough Mound to the base of Silbury Hill.† Thom had already clearly stated that the Neolithic builders were using whole numbers of megalithic yards or megalithic rods at Avebury, and these were commonly used in multiples of ten. (The distance on this line is 10,000 megalithic yards, or 4,000 megalithic rods, or 8.291 kilometers, which also equates closely to 4.5 nautical miles.)

---

*The Hatfield Mound at Marden Henge (now destroyed) was also, arguably, a second "partner" mound.
†A measurement of 8,284 meters appears precise without accounting for land slippage on Silbury Hill; 8,291 meters = 10,000 megalithic yards.

Figure 3.1. Marlborough Mound is now covered with trees. This is what it looked like in 1723 when William Stuckeley drew it.

Perhaps more importantly, GPS readings also revealed that a bearing of 90 degrees can be taken from one mound to the other. The precise line heading 90.00 degrees due east from the northern base of Silbury Hill extends to the southern base of the Marlborough Mound, accurate to about fifteen meters on the ground. But if the line is drawn from the same point at the base of Silbury Hill over the summit of the Marlborough Mound, it continues to the eastern landmass extreme of southeast England, the tip of the North Foreland Peninsula.

So there is considerable dependability in these Avebury alignments; joining two high points creates a line extending to a coastal extreme point. The same is found again by joining the Marlborough Mound summit to the summit of Leith Hill; this line extends to the tip of the Dungeness Peninsula, to the very same geometric point that is at right angles to Ness Point and Silbury Hill.

INCONCEIVABLE FACTS

In his book *From Carnac to Callanish,* archaeologist Aubrey Burl dismisses the idea that two Neolithic monuments could be aligned over a distance of fifteen kilometers. He finds it "inconceivable" that such an alignment was possible at the time.[3] But when all of the remarkable abilities of the human mind are considered, is it realistic to decide what is "inconceivable" and then attach that presumption to the entire population of Neolithic Europe? It cannot be sensible to insist that the first mathematical geniuses suddenly sprang up with the Greeks when, as Thom discovered, evidence to the contrary remains firmly planted in the bedrock of Wiltshire. This is not to say that the archaeological evidence of peasant living is flawed, but an examination of the detritus of a culture does not necessarily reveal the existence of, or the mental capabilities of, a minority of individuals. The monuments are perhaps a record of that ability. In society today the abilities of geniuses, savants, autistics, and remote viewers are recognized. Should we assume the same recognition did not exist five thousand years ago?

If Thom's beliefs were correct, individuals among the Neolithic people at Avebury were capable of communicating intelligent information through the *structure* of their monuments. With the benefit of technology that was not available to them, we can now suggest that the *precise location* of these monuments may help us to understand not only that they *could* do this, but also *why* they were doing it.

In their separate studies Alfred Watkins and Alexander Thom reached similar conclusions. Neolithic people made very accurate linear measurements over the land. The ability to align stones north to south with an accuracy of eight arc minutes (as Thom found at the Ring of Brodgar) or to align a pyramid within four arc minutes (as numerous surveys at Giza have concluded) can be weighed against the random chances of achieving this from among the 10,800 arc minutes available. And the tangential east-to-west alignment of Silbury Hill and Marlborough Mound is more accurate still.* The contrast between the orthodox and the alternative views of Neolithic ability could not be clearer. Thom and numerous other researchers extol the astonishing cleverness and accuracy of Neolithic measurements, while an orthodox vision finds no "meticulous positioning" and, flying in the face of Thom's analysis, concludes that stones were located in the Neolithic landscape following the principle that "near enough was good enough."[4]

Which of these views is correct?

IMMEDIATE EVIDENCE

There is immediate and striking evidence of highly advanced measurement in the Neolithic Period to be found in the location of the Avebury circle itself. A line of latitude running through the great circle center is 51.42857 degrees north of the equator. This precise number results from dividing 360 by 7. If 360 is divided by any other whole numbers from one to ten, the result is a whole number. The Avebury circle is located at this division with the accuracy of a guided missile. If this geodetic

---

*This is true if the line of latitude runs tangential to the circular base perimeter of both monuments, a method of Neolithic alignment recognized by Burl and others.

latitude was a choice rather than a chance, the builders communicate their ability to measure the planet extremely accurately; they also imply that if you are going to build your own hills in the landscape, you should line them up with natural extreme points in an orderly fashion.

Are these measures at Avebury simply a series of coincidences, or are they perhaps a faint echo of the ancient understanding of an ordered Earth? If so, can they give a direction to the evidence sustaining this belief?

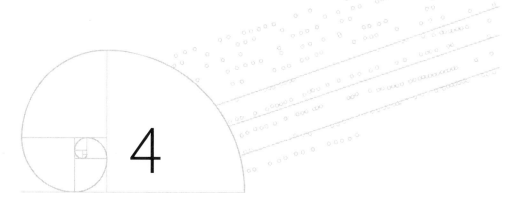

4

EARTH MEASUREMENTS

THE GEOID

Of all the scientific goals in human history, the pursuit of the Earth's true dimensions surely ranks among the most challenging. Sir Isaac Newton proposed that the Earth should bulge at the equator; the Italian astronomer Giovanni Domenico Cassini reckoned it should bulge at the poles. Nobody knew for sure who was correct, and scientific expeditions were dispatched to find the answer by measuring meridians in places as various as Lapland and Ecuador.*

After several years of trials and tribulations, the results were collated and Newton was proved correct. Measuring exactly one degree over the Earth on a meridian gives a greater distance at the poles than it does at the equator. The geoid is more like a grapefruit than a football.

This discovery presented a problem for humble travelers, dedicated astronomers, and intrepid explorers alike. Before this discovery navigators had been measuring in sea miles, assuming that the Earth was a perfect sphere. The circumference was divided into 360 equal parts called degrees, and these were each divided a further sixty times into minutes.

---

*Any line on the Earth joining the North and South Poles is a meridian, a line of longitude.

This sexagesimal system* gives a total of 21,600 minutes around the whole circumference of the Earth, and each one of these divisions was called a sea mile. After the Earth had been measured accurately, its modest equatorial bulge resulted in a sea mile at the equator being inconveniently shorter than one at the poles.† (The difference amounts to a maximum of about twenty meters in one arc minute.)

The problem was resolved with a compromise. The nautical mile was introduced because it is the mean value of one arc minute on the meridian. One nautical mile has an exact length: 1,852 meters.

This modern measurement is based in ancient history. The circumference of a circle may have been divided into 100 degrees or 850 degrees if another system had been adopted. Instead, the divisions of a modern circle are sexagesimal, and although its origins are unclear, this "base sixty" system was used at the dawn of recorded history in Mesopotamia. This ancient method of division, which results in 21,600 arc minutes around the Earth's circumference, gives us 21,600 nautical miles around the Earth's polar circumference today.

GETTING BEARINGS

The same ancient system is used to find modern bearings. To do this the circumference of the circle is again divided using the sexagesimal system, with zero degrees at true north. The navigator works clockwise from north measuring 360 equal intervals, or degrees, all the way around the circle. It's a brilliant system because any location on the globe can be specified from another location by using only these two measurements, a distance and a bearing.

From his study of ancient metrology, John Michell concluded that the slight difference between the Earth's equatorial and polar circumferences was known in prehistory. This, of course, is strongly contested,

---

*This is a "base sixty" system of counting. Hence, $6 \times 60 = 360$; $360 \times 60 = 21,600$.
†The two calculated distances are: one minute of arc at the equator, 1,842.9 meters, and one minute of arc at the poles, 1,861.7 meters.

but it is important to recognize that if these dimensions were known and a sexagesimal system was employed, then the nautical mile automatically results from recognizing them. The nautical mile is a logical unit to adopt when measuring long distances on the Earth because it is a length derived from the dimensions of the planet itself. If people living in the ancient world were somehow aware of these dimensions and used a sexagesimal system, they would no doubt measure bearings as we do today, and the same system would naturally produce the mean unit distance that we call the nautical mile.

A TRIBUTE TO RAND FLEM-ATH

In our early research at Carnac and Avebury, distances between monuments and topographical extreme points were measured in kilometers. This research ultimately revealed a common long-distance measurement adopted by the monument builders. This work was published on a website, with all the measurements given in kilometers. The author Rand Flem-Ath saw the data shortly after the website was launched, and he immediately recognized something intriguing. In a telephone call Flem-Ath asked, "Did you know that the common distance you have given in kilometers is almost exactly 1,618 nautical miles?"

It was Flem-Ath's insight that revealed the true sophistication of this Neolithic measurement. The digits 1618 are well known to mathematicians as the digits of phi. Flem-Ath said "Why not try pi as well?" Prior to this conversation we were unaware of the correlation. It subsequently became apparent that when the distances and bearings on lines between monuments and high points are measured using the sexagesimal system, evidence of its use is abundant. A common system was adopted by these monument builders, and the reason becomes clear; they were describing an ordered Earth that is still visible today. The largest and most famous of the monuments provide evidence of this.

England's second-largest stone circle monument, at Stanton Drew, provides an excellent example of the Neolithic ability to measure their landscape; moreover, the choice of location for the circles at Stanton

Drew clearly indicates that the alignment of high points found at Avebury was not simply an accident.

STANTON DREW

Stanton Drew is located in a valley about fifty kilometers west of Avebury. On a bank alongside a country lane, the garden at The Druid's Arms in Stanton Drew has two standing stones with a fallen, or recumbent, stone between them. Legend has it that the two uprights are the bride and groom and the recumbent third is the presiding priest, possibly the earliest example in history of "getting horizontal" at a wedding. A short distance away there are circles of dancers and an avenue of fiddlers who were lured into forgetfulness by the devil. They were all turned to stone when the riotous festivities of a Saturday continued into the early hours of Sunday. It must have been quite a night. The bridal party and guests—three contiguous circles, with many still upright—make this the second largest of England's Neolithic stone circle monuments after Avebury. The location of Stanton Drew, in the lowland of the Chew Valley, does not immediately draw comparison with Avebury. Until, that is, lines are drawn from the monument to the regional high points.

CLEEVE HILL AND PEN Y FAN

The regional high points to the northeast and northwest of Stanton Drew are Cleeve Hill and Pen y Fan.* Cleeve Hill is the highest summit in the Cotswold Hills, and Pen y Fan is the highest summit in South Wales.

The two lines drawn from the regional high points to Stanton Drew are shown in figure 4.1, but both lines have been extended to the coast, just as they were at Avebury. In both cases at Stanton Drew these lines extend to southern landmass extreme points.

The two extended lines in figure 4.1 pinpoint two southern

---

*Cleeve Hill and Pen y Fan are the highest points to the northeast and northwest of Stanton Drew south of the fifty-second parallel.

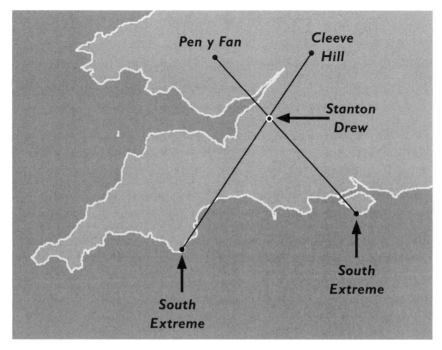

Figure 4.1. Stanton Drew in the crosshairs of lines
joining topographical extremes.

landmass extreme points: the southern limit of the Isle of Wight at
St. Catherine's Head and the southern limit of Devon at Prawle Point.

The intersecting lines produce two odd measurements:

1. A bearing of 314.27 degrees can be taken from Stanton Drew to
   Pen y Fan's summit.
2. A bearing of 31.42 degrees can be taken from Prawle Point, over
   Stanton Drew's largest circle, to Cleeve Hill's summit.

The process of alignment at Stanton Drew duplicates that recognized at Avebury; both monuments are in the crosshairs where lines
joining regional high points and coastal extreme points intersect. It
would appear that, consciously or otherwise, the monument builders
have communicated a relationship between high points and landmass
extreme points in nature.

The line from the tip of Prawle Point cuts through Stanton Drew's largest circle and continues to the summit of Cleeve Hill. This line has a bearing of 31.42 degrees and passes to a summit point about sixty meters from the Ordnance Survey's concrete trig point. Measuring to the same summit point from the Stanton Drew circle gives a bearing of 32.36 degrees: twice 16.18.

Were the builders operating on a level unimagined in the present day? If so, they appear to have been identifying measured relationships in the landscape—and doing so in a systematic manner. The monument location identifies four distinguished extreme points in the landscape by aligning between them. Why would the monument builders do this? The answer is found by joining the specified natural points *independent of the monument*. When this is done a remarkable natural landscape geometry emerges. Joining these points at Stanton Drew provides an illustration of this.

The four topographical extreme points specified by their alignment with Stanton Drew are joined with each other independently. Each resulting line suggests order in the landscape.

1. The line joining St. Catherine's Head with Cleeve Hill's summit (trig point) extends to the summit of Britain's highest mountain, Ben Nevis. Hence, a line is revealed joining the national high point, a regional high point, and a landmass cardinal extreme point: ^-^-^.

2. The line joining Cleeve Hill's summit with the summit of Pen y Fan continues to Ireland, where it cuts across the southern peninsula of Ireland's landmass at Mizen Head.* Hence, the alignment of topographical extremes is repeated: ^-^-^.

3. The line joining Prawle Point and Pen y Fan continues to the tip of Fife Ness, a pronounced chevron-shaped peninsula on the east coast of Scotland; again: ^-^-^.

---

*The line from Cleeve Hill over Pen y Fan cuts across the peninsula with Mizen Head and Brow Head on it. If the line is taken to the southern tip of the headland, it goes from Cleeve Hill summit over the base of Pen y Fan.

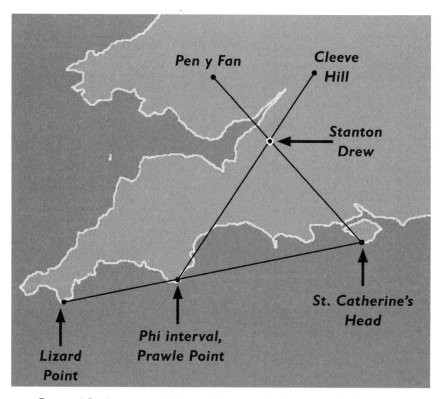

Figure 4.2. A repeat of figure 4.1, showing Stanton Drew in the crosshairs, but this time the two points on the south coast are joined and the line is extended to the west. The base of Prawle Point is found at the phi interval on the line.

4. The line joining the two southern coastal extremes is the most illuminating. It passes from St. Catherine's Head over the base of the Prawle Point peninsula and continues to mainland Britain's most southerly coastal rocks, at Lizard Point, also known as the Lizard. Thus, these three southern landmass extremes of Britain are found in alignment: ^-^-^. Also, as shown in figure 4.2, this line is divided at the phi interval by Prawle Point.

The next image, figure 4.3, shows that the phi interval is located on the beach at the base of Prawle Point, about six hundred meters from the tip.

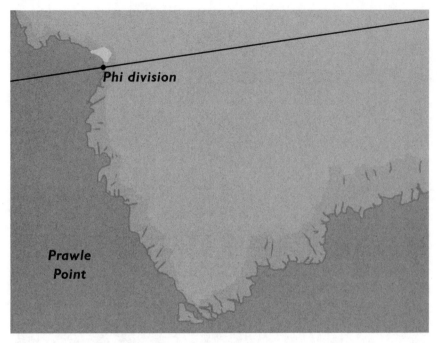

Figure 4.3. Prawle Point and the phi interval.

Having joined all four of the specified topographical points, the lines have extended to:

1. The tip of a large chevron-shaped peninsula in Scotland.
2. The southern peninsula of the British mainland.
3. The southern peninsula of the Irish mainland.
4. The highest summit overlooking both landmasses.

So by following this process at Stanton Drew, the monument appears to draw attention to a series of natural three-point alignments across Britain, with one of these lines expressing the phi proportion.

THE NATURE OF PHI

Phi was referred to as "the divine proportion" in the Renaissance; it could also be called nature's favorite ratio. The proportion of 1 to 0.618 is found

Figure 4.4. Phi Ratio. The ratio of AC to BC
is the same as the ratio of BC to AB.

in beehives, seashells, sunflowers, pinecones, plant stalks, insect segments, and on and on. It is even in our wallets. Multiply the length of the short side of a credit card by 1.618 to get the length of the long side.

Phi is the most magical and mysterious of all numbers. The digits are 1.618033 . . . , and like pi, they go on forever. Euclid called phi "the golden mean." It has many unique mathematical properties.

The phi ratio, or something very close to it, appears so frequently in nature it is hard to find anywhere where its presence is entirely absent. It has most recently popped up in the DNA spiral; the width-to-length ratio of each spiral is said to be 1:1.619.

It seems rather surprising, and not a little illogical, that although the phi proportion is the foundation of much that is beautiful in the natural world, it is not a proportion commonly associated with topography and the landscape. Nevertheless, the phi interval on Britain's south coast has been recognized by following a simple linear progression from a Neolithic monument. What emerges is a view of geography not commonly recognized today: a landscape where a relationship between cardinal coastal extreme points and high points is defined by geometry and number.

CROSS-COUNTRY COMMUNICATION

The seemingly random location of Stanton Drew is ultimately revealed as an extraordinary magnet for lines joining high points and coastal extreme points. And by following this process of ^-^-^ alignment for a second time, the result once again provides insight into an underlying order in the landscape. A line from Stanton Drew to the highest summit in Wales (and all southern Britain), Mount Snowdon, extends to the eastern extreme of Ireland's landmass. In turn a line from Stanton Drew to the highest summit in Ireland, Carrantuohill, extends to Inishvickillane, which is the last in a chain of small islands, just off the Dingle Peninsula, on the western extreme of Ireland.* This long finger of land terminates in a smaller, thinner finger that supports Ireland's most westerly flock of sheep. Figure 4.5 shows these two lines from Stanton Drew.

Both of the lines shown in figure 4.5 read ^-^-^, with each line joining the monument with a national high point and a cardinal extreme point. These additional alignments single out the location of Stanton Drew as one with unique qualities. The degree of knowledge and ability required

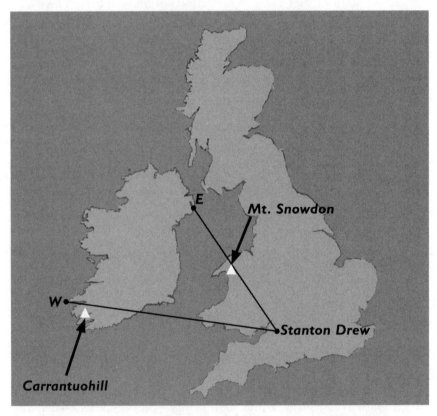

Figure 4.5. Stanton Drew and the alignment of national high points with the eastern and western extremes of Ireland.

---

*The small island of Tearaght, not in the chain, is farther west, but not specified by the monument.

to align a monument *intentionally* with high points and topographical extreme points in this sophisticated manner greatly exceeds the generally agreed scope of Neolithic ability. How could they possibly have known where these extreme points were? And were the extreme points in the same place when the monuments were built? These questions tend to negate the value of any further research in this area. On the other hand, if these alignments between the monuments, high points, and extreme points are purely accidental, why do the same crosshair alignments occur at the two largest Neolithic circles in the country? The conclusive evidence that the monument builders were specifying these remote points intentionally is found by joining them together *independently* with straight lines. When these points are joined, independent of the monument that specifies them, the resulting measurements are invariably informative; they all describe a natural pattern in the landscape. Ireland's western and eastern extremes are therefore joined together to reveal this pattern.

A LINE OVER IRELAND

Ireland's highest summit, Carrantuohill, is located among a dramatic cluster of mountains called the MacGillycuddy's Reeks on the southwest of the landmass. With the exception of the mountains in this group, the next highest point in Ireland is Mount Brandon, located to the north of the town of Dingle, on the Dingle Peninsula.

When the eastern and western extremes of Ireland (as specified by alignment with Stanton Drew) are joined together, the resulting line passes over the summit of Mount Brandon. Similarly, when England's eastern and western extremes (specified by alignment with Avebury) are joined, the line passes over High Willhays. Therefore, by joining the specified points at Avebury and then at Stanton Drew, both England and Ireland are found to have notable regional summits on the line joining their eastern and western extreme points. Both monuments act in the same way: they draw attention to this mysterious duplication on two neighboring landmasses, an aspect of topography not generally recognized today.

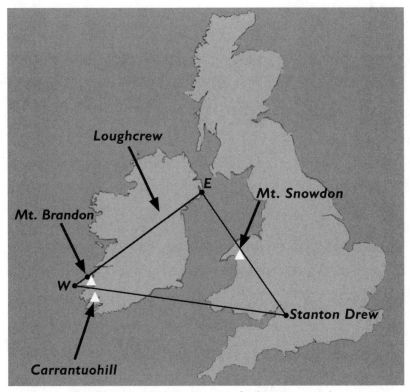

Figure 4.6. The line joining the specified eastern and western limits of Ireland passes over Mount Brandon and Loughcrew. Both extreme points (W and E) are highlighted by lines from Stanton Drew passing over the highest summits in Wales and Ireland.

The significance of this cross-country line for Neolithic Ireland is evident because it has one of Ireland's largest Neolithic monuments located on it, the cairns at Loughcrew. The alignment of Loughcrew between the eastern and western extremes of Ireland duplicates the alignment of Avebury between the eastern and western extremes of England (figure 4.6). Evidence of a sophisticated mutual understanding can be measured on the Irish line. The highest hill at Loughcrew, Slieve na Calliagh, has numerous enormous Neolithic cairns on its summit; from the base of this hill, 161.8 nautical miles can be measured along this line to the western extreme at Inishvickillane.

AN ALIGNMENT PRINCIPLE

An alignment principle was suggested at Carnac, but it appears to be spelled out at Avebury and replicated at Stanton Drew and Loughcrew. The monuments describe an ordered landscape. This measuring process results in the discovery that the distance between the western landmass extremes of Ireland and the U.K. mainland is 314.22 nautical miles when measured in ground length on Google Earth. Perhaps the greatest barrier to learning more about this Neolithic method is the deep-seated impression that measurements of this magnitude and complexity were impossible at the time. But however deep-seated it may be, this is only a present-day impression. The extent of these people's mental abilities cannot be known, so we make a guess. From the preserved corpse of the man called Ötzi from 3000 BCE, which was found frozen in a glacier in the Alps, we have learned what he ate, what he wore, and the items he considered useful; it is even known how he was killed. But his thoughts, skills, and aspirations are entirely unknown. Great minds are not revealed through the study of empty skulls. And if these measurements were not possible at the time, we need to ask ourselves what else could explain not only the number of seemingly sophisticated alignments, but also their replication at three of Europe's largest Neolithic sites.

If this alignment process at Britain's two largest Neolithic circles was intended, it should also be evident at the third largest site.

BRODGAR

The circle at Brodgar on the island of Mainland, the largest island in Orkney, was evidently created with geometry and sexagesimal measurements in mind. Brodgar was built as a perfect circle about one hundred meters in diameter on the western part of Mainland. The circle of sixty equally spaced stones is surrounded by a henge that was dug into the bedrock.

Figure 4.7. The circle at Brodgar.
Photo by Hans Peter Schaefer.

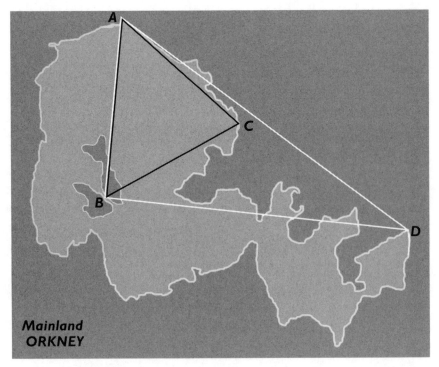

Figure 4.8. The geometry at Brodgar, marked B in the figure.

1. A bearing of 6.18 degrees can be taken from B to A at the northern limit of the landmass of Mainland.
2. A bearing of 61.8 degrees can be taken from B to C at the eastern coastal limit shown in figure 4.8.
3. A distance of 16.18 nautical miles can be measured from B to D, the tip of the chevron-shaped landmass in the east.*

These three natural extreme points are specified by phi-digit measurements taken from the circle. The sophisticated nature of this process is again revealed by joining the specified points together. The result is a right triangle and an isosceles triangle within it, with both triangles having corners at Brodgar (see figure 4.8). It is possible to draw the isosceles triangle with legs measuring 8.09 nautical miles, so 16.18 nautical miles results from adding them together. This can be done using a geometric point at the base of the northern peninsula.

The circle at Brodgar is therefore a geometric point in a landscape where extreme topographical points are also treated as geometric points.

The highest summit to the north of the Brodgar circle is Mid Tooin. A line from this summit to the circle extends to the western extreme of Orkney, a point near Rora Head. A distance of ten nautical miles can be measured on this line from the circle to the western landmass extreme.

Each of the four monuments discussed so far is aligned between a regional high point and a coastal extreme point. It seems unlikely that the similarity between the alignments at Brodgar and those at Avebury, Stanton Drew, and Loughcrew could arise from wandering aimlessly around Britain and stopping four times at random. However, it seems equally unlikely that measurements of such a precise nature could have been made in the Neolithic. If these alignments and the associated measurements were devised intentionally, they would have required an aptitude not generally believed possible today. Moreover, in all four cases

---

*The distance of 16.18 nautical miles measured from the circle's eastern extreme arrives at the flattened tip of this chevron of land.

the monuments are aligned with static regional high points and extreme points on the coast that are subject to varying degrees of change over the millennia. For this reason coincidence may appear to be the only reasonable explanation for the repetition of alignments. But is coincidence a sensible explanation?

If the coastal extreme points were not recognized as geometric points by the monument builders, it must be imagined that any random point in the landscape will currently align between a regional high point and a coastal extreme point. Although this is not true, it could still be argued that there are lots of extreme points and alignments like these will always be found eventually. But even if this were true, it would not provide an explanation. The evidence of a common code of practice does not depend on the alignments alone; rather it depends on the repetition of pi and phi digits when measuring the lines. This repetition is like a common signature that undermines, if not eliminates, pure chance as a coherent explanation. In other words a pattern becomes predictable. The points that are aligned with the monuments can be joined together to form geometric shapes—right and isosceles triangles—and when this geometry is measured, there are a disproportionate number of pi and phi digits in the results. It is this predictability, from one monument to the next, that indicates the nonrandom nature of these monument locations.

If this truly is a repeating pattern, it should be in evidence at Stonehenge, Britain's most famous Neolithic monument.

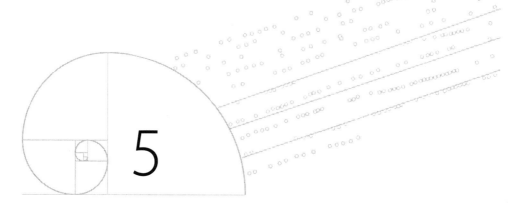

5

Circular Reasoning

STONEHENGE, TWENTY-EIGHT KILOMETERS SOUTH OF AVEBURY

By around 3000 BCE work on the monuments at Avebury, Stonehenge, and Stanton Drew had begun. Some four hundred years later, Silbury Hill was built. Stonehenge was eventually completed around 2100 BCE. For the Neolithic people it would seem that monument building was about the process as well as the product. Life spans of thirty-five to forty years were normal at the time, and many generations were born, labored, and died before the final result was achieved. The monuments made an indelible mark on the landscape, defining a particular location for thousands of years to come.

In 1966 a new parking area was built at Britain's most famous Neolithic monument to accommodate the ever-increasing number of visitors to the site. Mechanical diggers began leveling an area about three hundred meters from the Stonehenge circle, and during the excavation three large ancient postholes were uncovered. Charcoal taken from the holes revealed that they were dug into the bedrock between 8000 and 6700 BCE, several thousand years before the famous stones were erected.

The ancient postholes were marked with white paint in the parking area.* It is strange to look down on these circles at your feet and realize that someone many thousands of years ago also stood here looking at the ground. A turn of the head, to gaze up a slight bank, and there loom the trilithons of Stonehenge with their huge upright stones breaking the bleak, tilted horizon of the Salisbury Plain. The attraction of this seemingly arbitrary location may date back fully ten thousand years to the time when the first large posts were erected. Several thousand years later the first large circle was marked out at the site, with fifty-six holes around its circumference. Enclosing this ring, a near-circular henge was cut through the thin layer of top soil into the chalk bedrock. At intervals over the next nine hundred years, wooden and stone circles were built and rebuilt within the henge bank.†

Many stones used in the construction of both Avebury and Stonehenge are sarsen, an extremely hard sandstone, a mixture of sand and silica cement, and favored by builders down the years for its durability.

The thirty upright stones in the sarsen circle at Stonehenge each weighed about twenty-five tons. They offer no hint as to why anyone would wish to drag them to this bleak location and stand them upright. Such a thing is scarcely imaginable from the comforts of our twenty-first-century living. In their "primitive" conditions the builders must have suffered illness and perhaps hunger; with their short life spans, they would surely have had more urgent considerations than to stand these stones upright on unlevel ground and surmount them with a further thirty stone lintels, cut in curves and carefully jointed together. The result was a horizontal stone ring weighing over one hundred tons suspended over four meters above the ground.

THE SOLSTICE LINE

Through its long history Stonehenge has seen it all—and been it all. Some surmise it was a marketplace; others the site of gruesome human

---

*As of 2013 the entire area is being relandscaped.

†Nearly all dates for the successive works at Stonehenge are subject to some disagreement. The dates given here are ballpark figures.

sacrifices, an astronomical observatory, a sanctuary, a temple, or simply a place to gather. Every Midsummer's Eve the stones still exert their influence. People from all walks of life—Druids, Pagans, Christians, celebrities, novelists, artists, and artisans—all meet here. Well-heeled gentry with their 4×4s and picnic hampers, the businessmen, the bikers, the hikers, the hippies, the new agers, and old-ageing flower children. What have they come to see? They have gathered to participate in a great British tradition, to watch the sun rise at the summer solstice at Stonehenge.

The summer solstice is the single day of the year when the sun rises at its northern extreme on the horizon. Viewed from a central point in the stone circle, the sun's rising orb passes over the enigmatically isolated Heel Stone some eighty meters away. Whether this solar alignment was intended is a moot point. The bearing to the midsummer sunrise has altered by more than one degree since the last stones at Stonehenge were erected. This moving solar target attracted the attention of an American, James Jacobs, host of an Archaeogeodesy website,* who, like Thom and many others, has discovered surprisingly accurate measurements among the Neolithic monuments. Jacobs believes that the intended bearing to the Heel Stone may have been 51.4285 degrees (the same figure as the latitude of Avebury); Thom suggested 49.95 degrees. Either could be correct, or perhaps both. Excavations suggest that the Heel Stone possibly had a partner, long since destroyed.

If, as the evidence begins to suggest, the Neolithic people embraced landscape geometry, how good were they? Could they really calculate bearings so accurately?

JACOBS'S LEAD

Jacobs suggests that measurements of this accuracy may have been possible, so his lead was followed. A bearing of 51.4285 degrees was taken

---

*Jacobs's website can be found at www.jqjacobs.net (accessed June 27, 2015).

from Stonehenge, and the line was drawn over the landscape. This line goes directly to the highest regional point, Walbury Hill, passing to the Ordnance Survey "viewpoint" on the northern limit of the hilltop earthwork. In this way the alignment signifies the use of geometric points on Walbury Hill in the Stonehenge geometry, and sure enough a line drawn from this northern limit of Walbury Hill passing back over Stonehenge continues to the tip of Prawle Point.

Thus, the Stonehenge line reads ^-^-^ by aligning the monument between a regional high point and a landmass extreme point. Both of these two topographical extremes, Prawle Point and Walbury Hill, have already been identified by alignments, one with Avebury and one with Stanton Drew. If we are beginning to grasp the language of the monuments, we should expect a crosshair alignment at Stonehenge in keeping with those at the other two monuments. There is one, and it includes the highest summit on the landmass. A line from Ben Nevis passing over Stonehenge extends southward to the western landmass extreme of the Isle of Wight (see figure 5.1).

If these alignments with the regional and the national high points

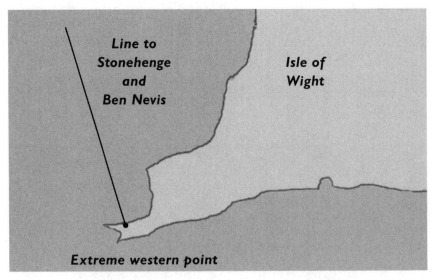

Figure 5.1. Tennyson Down on the western landmass
extreme of the Isle of Wight (overlooking The Needles);
the line passes over Stonehenge and Ben Nevis.

were intended, then two landmass extreme points in the south have now been specified by two ^-^-^ alignments with Stonehenge, first Prawle Point and then Tennyson Down.

The probability that this is simply a coincidence can be judged by following the procedure adopted at the other monuments. At Avebury two coastal extremes identified by hilltop alignment created a right triangle with Silbury Hill; the same thing happens at Stonehenge. A right triangle can be drawn between Tennyson Down, Prawle Point, and Stonehenge, as shown in figure 5.2.

This Neolithic geometry invariably produces pi and phi digits when the lines are measured. A line from Stonehenge to the base of Tennyson Down has a bearing 161.8 degrees. A distance of 2 × 16.18 nautical miles can be measured between these points. This line from Stonehenge to Tennyson Down provides a baseline for the right triangle shown in

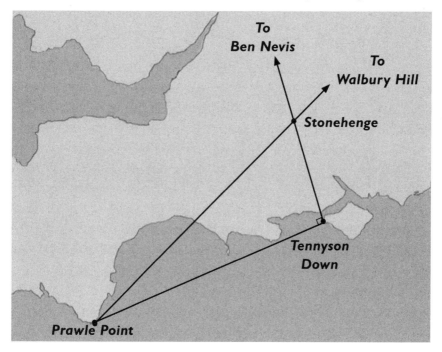

Figure 5.2. Stonehenge in the crosshairs. A right triangle can be drawn between Stonehenge, Tennyson Down, and Prawle Point. The arrows indicate the alignments of two triangle sides with Ben Nevis and Walbury Hill summits.

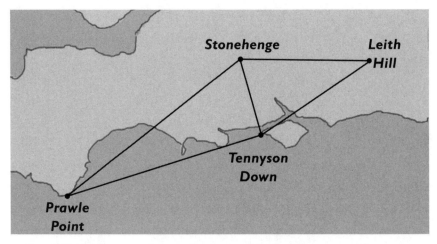

Figure 5.3. The joined isosceles and right triangles created between Stonehenge, Tennyson Down, Prawle Point, and Leith Hill.

figure 5.2, but it is also the baseline for an isosceles triangle with the now-familiar regional high point of Leith Hill at the apex. Figure 5.3 shows the joined isosceles and right triangles.

A COMMON PROCEDURE

Pi- and phi-digit measurements once again appear with great frequency. The distance from Stonehenge to the northern landmass extreme of the Isle of Wight is 31.42 nautical miles. Again, this measurement accords with the alignment of Stanton Drew, Pen y Fan, and the southern extreme of the Isle of Wight. So three cardinal landmass extremes on the Isle of Wight have each been specified with pi- and phi-digit measurements from the monuments. To discover why these natural extreme points are specified in this way, they are joined together independently. The result is an equilateral triangle (see figure 5.4).

The natural geometry shown in figure 5.4 is completed in a remarkable way when the highest point on the Isle of Wight, at St. Boniface Down, is added as a geometric point. A right triangle is created. A geometric motif emerges in nature by joining the extreme topographical points on the island; joined isosceles and right triangles (figure 5.5).

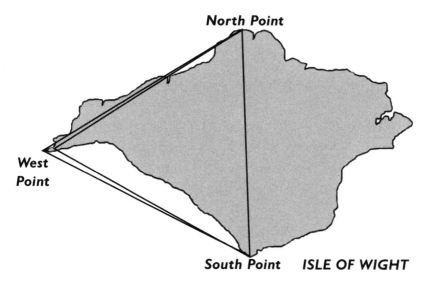

Figure 5.4. The Isle of Wight. Three cardinal extremes of the island form an isosceles triangle when joined together. But if the western point is moved to the base of the peninsula at Tennyson Down, an equilateral triangle is formed. This base area is located from Stonehenge on a line bearing 161.8 degrees, which is 2 × 16.18 nautical miles in length.

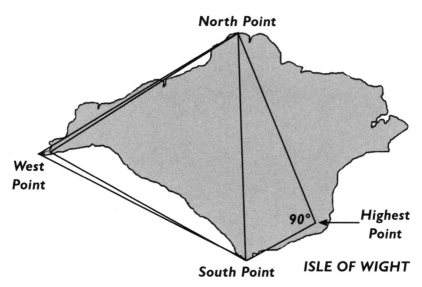

Figure 5.5. The Isle of Wight, illustrating the joined isosceles, equilateral, and right triangles created between three cardinal coastal extremes of the landmass and its highest summit.

LANDSCAPE GEOMETRY

From the island's western extreme, a line bearing 6.18 degrees extends to the summit of Walbury Hill. Here the geometric motif created between the natural extreme points is repeated among the monuments. The shortest distance from the Ordnance Survey viewpoint on Walbury Hill to Avebury Henge is fifteen nautical miles. Exactly the same distance can be measured from this point on the Avebury Henge to Stonehenge.

The result is an isosceles triangle drawn as the crow flies between the two famous Neolithic monuments and the viewpoint on the highest regional summit that overlooks both of them. The monuments are therefore creating geometry by adopting high points and extreme points as geometric points, and the evidence that this was intentional becomes increasingly clear with a second topographical triangle.

Again with Stonehenge and Avebury at two corners, a right triangle can be drawn with Glastonbury Tor at the third corner (figure 5.6).

The geometry between the four points suggests that the henges at both Stonehenge and Avebury were intended to define geometric points in the landscape. The precise geometric points are found

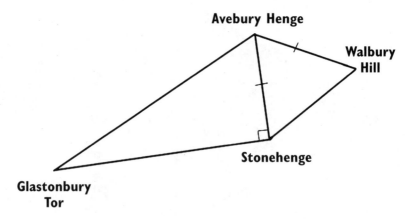

Figure 5.6. A right triangle and an isosceles triangle share a common line that joins Avebury Henge and Stonehenge. The two other geometric points are Glastonbury Tor and the highest summit in the region, Walbury Hill.

within the orbit of each henge bank. In the case of Walbury Hill, the precise point is found at the viewpoint defined by the Ordnance Survey on the northern limit of the earthwork on the flat summit, but in the case of Glastonbury Tor, the geometric point is found at the base of the hill.

GLASTONBURY TOR

Seen from thirty kilometers away, Glastonbury Tor rises from the Somerset Levels like a small incongruous volcano. Spiral earthworks wrap around the base of the hill and up to the summit like a coiled snake. It is a curiously enormous hill, entirely reshaped by human endeavor.

The unusual relationships between regional high points and the monuments at Brodgar, Stanton Drew, Loughcrew, Stonehenge, and Avebury offer a clear indication that the monuments were carefully located with landscape geometry in mind. The phi- and pi-digit measurements indicate that the Neolithic motivation cannot be understood without recognizing the profoundly advanced megalithic science already recognized by Thom and numerous other researchers.

A GEOMETRIC MOTIF

It is undoubtedly true in an investigation such as this that if enough targets are available, a dedicated searcher will find something that fits. But equally, it is not true to say that the same geometric motif will arise in all cases, nor is it true that duplicate measurements will occur in all cases. This is, however, true when a broad study of the stone circles in southwest England is conducted. Two images from Cornwall, figures 5.7 and 5.8, provide examples of how the same geometric motif is repeatedly created by joining a stone circle with surrounding topographical extremes. It is the near constant use of this motif that identifies the common code of practice at work; the specified points in the geometry are always on the highest points or on the cardinal extreme point of a

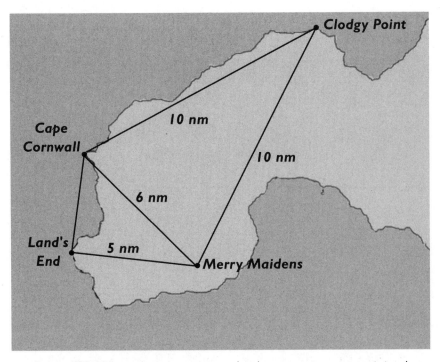

Figure 5.7. When the two western landmass extremes are joined with the Merry Maidens stone circle, the result is a right triangle. The hypotenuse of the right triangle forms the baseline for an isosceles triangle with the northern landmass extreme at the apex.

landmass. The first image, figure 5.7, shows the location of Cornwall's Merry Maidens stone circle creating an isosceles triangle and a right triangle with landmass extremes.

The hypothesis that nautical miles were adopted by the people who chose this location appears self-evident when the lines creating this geometry are measured accurately.*[1] In addition to the measurements shown above, a distance of 16.18 nautical miles can be measured from the Merry Maidens circle to England's southern extreme, among the southern rocks on the Lizard. The great majority of stone circles in Cornwall conform to this pattern. One further example of this geom-

---

*These lines were measured using the Vincenty formula, which accounts for the slightly non-sperical shape of the Earth.

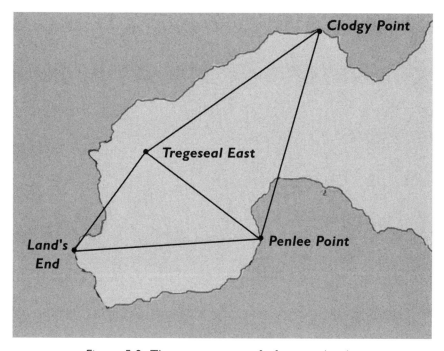

Figure 5.8. The common motif of a joined right triangle and an isosceles triangle is created with lines drawn between topographical extreme points and the Tregeseal East stone circle in Cornwall.

etry is shown in figure 5.8, this time with measurements taken from the Tregeseal East stone circle.

These circles all appear to be carefully placed in the landscape to produce the same geometric motif found at Avebury, Stonehenge, and Brodgar. They repeatedly adopt the cardinal extreme points of a landmass as geometric points. And when these natural extremes in Cornwall are joined together independently (as they were on the Isle of Wight), they too are found to be mysteriously geometrically ordered. So although coastlines may have altered through the ages, they have currently worn themselves into an ordered geometric arrangement on the Isle of Wight and in Cornwall, as shown in figure 5.9.

The world-renowned St. Michael's Mount is just visible in the far right of figure 5.10, on the small island. This landmark is also specified

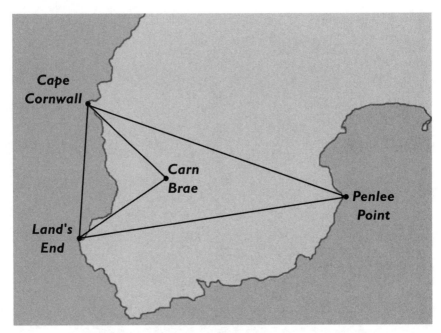

Figure 5.9. An isosceles triangle can be drawn between
the two western and the eastern extreme points of
the peninsula. Within this triangle the highest summit
in the region, Carn Brae, is also equidistant from the
peninsula tips at Cape Cornwall and Land's End.

by the Cornish monuments. The motif of a joined isosceles triangle and
a right triangle appears again among the four natural extreme points
when the summit of St. Michael's is adopted as a geometric point. More
detailed work on Cornish circles is found in appendix 1.

Pi- and phi-digit measurements recur with great regularity when
measuring between the natural extreme points specified by the monuments. When these points in Cornwall are measured, the bearing from
the southern extreme at Hella Point to the northern extreme at Clodgy
Point is 31.42 degrees. And again the bearing from the southern
extreme of the Lizard to Cornwall's highest point, Brown Willy, is also
31.42 degrees. Moving up the south coast to Prawle Point, a bearing of
31.42 degrees finds the Cotswold's highest point, Cleeve Hill. Move to
St. Catherine's Point, and it is 314.2 degrees to South Wales's highest

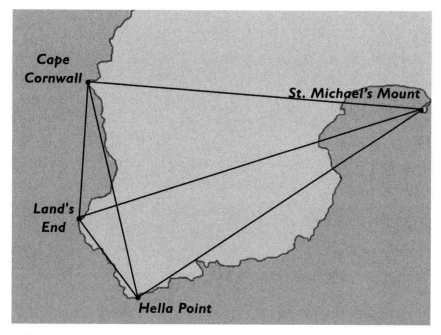

Figure 5.10. Joined isosceles and right triangles are created by joining the summit of St. Michael's Mount with three coastal extreme points.

point, Pen Y Fan; or move to the tip of Tennyson Down, and it is 6.18 degrees to Walbury Hill. All these points are specified by their alignment with stone circles. The Neolithic builders are defining an esoteric order on the landmass.*

People all those years ago were measuring the Earth, and whoever they were they saw it from an entirely different viewpoint.

---

*At the risk of being tedious, the list doesn't stop there: a bearing of 31.42 degrees can be taken from a point on the Dingle Peninsula to Ben Nevis; and the same bearing can be taken from St. David's Head to Mount Snowdon.

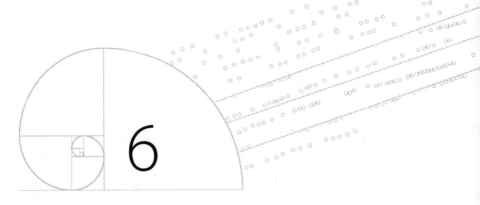

6

MOUNTAINS AND
MONUMENTS

THE ROLLRIGHT STONES, FIFTY-FIVE
KILOMETERS TO THE NORTH OF AVEBURY

Just off a side road, an inconspicuous little parking area among the trees causes drivers to wonder if they have pulled into the right place. But through the undergrowth the ancient King's Men stone circle at Rollright comes into view. Pockmarked, or "moth-eaten," as the antiquarian Rev. Dr. William Stuckley described them, these old and timeworn stones promise fertility to the woman who presses her breasts against them. We cannot verify this, having seen no bare-chested women on our visits, although stormy weather and biting winds may have curbed their enthusiasm.

In 3000 BCE, if not earlier, the site was marked with five upright stones, known as the Whispering Knights.* Whoever found this location was, it seems, a master seer or surveyor, a draftsman who drew lines to extreme points in the landscape.

---

*The Whispering Knights appear to be part of a destroyed dolmen or part of a long barrow.

About three hundred kilometers to the north of Rollright are the two highest summits in England, Scarfell Pike and Scarfell; the two boulder-strewn summits are only about one kilometer apart. A line from the Rollright circle passes between these two summits and extends to the highest summit on the entire landmass, Ben Nevis.

Britain's second highest summit is Ben Macdui, about eighty-five kilometers from Ben Nevis.

1. The distance from the Rollright circle to the summit of Ben Nevis is 313.69 nautical miles.
2. The distance from the Rollright circle to the summit of Ben Macdui is 314.86 nautical miles.
3. The mean distance from the Rollright circle to the two highest summits in Britain is 314.27 nautical miles.[*1]

It is therefore possible to draw an isosceles triangle with legs measuring 314.2 nautical miles and with the Rollright circle at the apex. The two legs of the triangle extend to the two highest mountains on the landmass, that is, the precise corners of the triangle are found on the bodies of these two mountains. A right triangle can be attached to the isosceles when a geometric point is placed on Lundy Island, in the Bristol Channel.[†2] The two triangles joining topographical extremes with Rollright are shown in figure 6.1.

Are the builders identifying themselves through this geometry with pi-digit measurements? The line from Rollright to Ben Nevis passes between the two highest summits in England, Scarfell Pike and Scarfell. These summits are one kilometer apart, and from a point on their northern base, it is 161.8 nautical miles back to the Rollright circle.

In the cases of Avebury, Stonehenge, Stanton Drew, Loughcrew, and Brodgar, lines to high points extend to coastal extreme points. The same applies at Rollright. The straight line from Ben Nevis passes

---

[*]These lines were measured using the Vincenty formula.
[†]Lundy Island was also adopted as a geometric point by the Stonehenge builders.

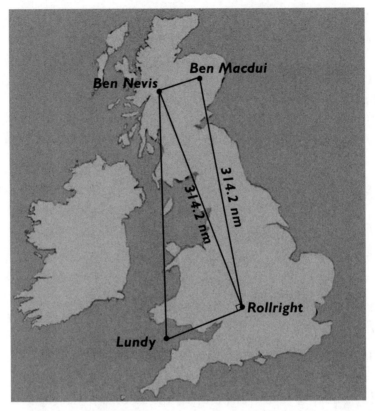

Figure 6.1. Joined right and isosceles triangles are formed with lines drawn between Rollright, the two national high points of Ben Nevis and Ben Macdui, and Lundy Island.

between England's two highest summits and then continues over the Rollright circle. It extends southward to the flank of Selsey Bill, which, not unlike Dungeness, is a large chevron-shaped peninsula jutting into the sea on England's south coast. From the tip of Selsey Bill at low tide, a distance of one-half of 161.8 nautical miles can be measured to the King Stone at Rollright.

It is the repetition of the same code of practice from one monument to the next that makes its common use ever clearer. In all cases the same procedure identifies the extreme topographical points of Britain as geometric points; in turn, by joining these points the landmass itself can be seen as a geometrically ordered object.

LANDSCAPE DESIGN

The overall result of studying these major Neolithic monuments is to discover a repetitive reference, through alignment and measurement, to the principal topographical extreme points in Britain and Ireland.

By following these alignments from one stone circle to the next, we are drawn into a world very different from our own. It is a calculated world, and the evidence of this ordered landscape is located at the corners of the Earth, the cardinal extreme points of landmasses.

When the outer extremes of Britain's landmass are joined together, a formal pattern emerges; the corners of the landmass appear to be located systematically. In the sequence of the five images of Britain that follow (see figures 6.2 to 6.6), the isosceles triangle apex is moved from one landmass extreme to the next, starting with Ness Point in the east, then the northern tip of Cape Wrath, then the northern tip of Dunnet Head, then the northern limit of Shetland, and then the tip of Peterhead.

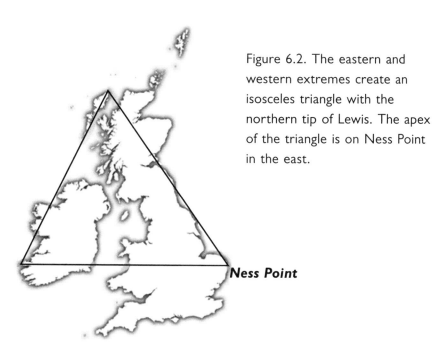

Figure 6.2. The eastern and western extremes create an isosceles triangle with the northern tip of Lewis. The apex of the triangle is on Ness Point in the east.

Ness Point

Figure 6.3. The eastern and western extremes create a second isosceles triangle, this time with Cape Wrath at the apex.

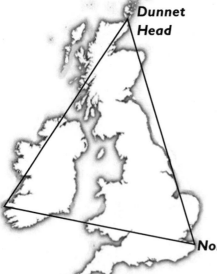

Figure 6.4. These western, eastern, and northern extreme points form a third isosceles triangle. Dunnet Head is at the apex, and the triangle corner in the east is on the tip of North Foreland.

Shetland

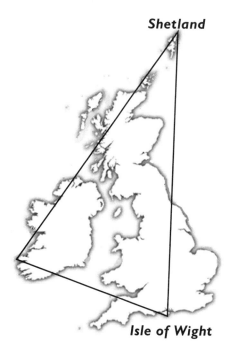

Isle of Wight

Figure 6.5. The northern limit of Shetland and the southern limit of the Isle of Wight also create an isosceles triangle with the western extreme. This is the longest of the triangles, and the legs measure 618 nautical miles to an apex on the rocks on the northern extreme of Shetland.

Peterhead

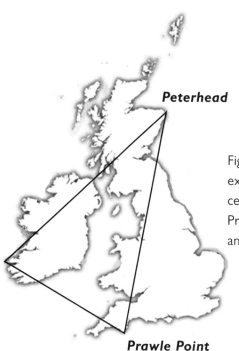

Prawle Point

Figure 6.6. The western extreme forms a fifth isosceles triangle, this time with Prawle Point on one corner and Peterhead at the apex.

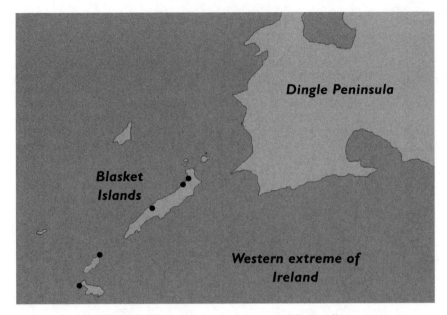

Figure 6.7. The Blasket Island chain is on Ireland's western extreme, just off the western extreme of the Dingle Peninsula. The five dots indicate the location of the corners of the five isosceles triangles shown in figures 6.2–6.6. The length of the island chain is about ten kilometers.

These five triangles give the impression of a systematic ordering of the landmass extreme points because each triangle has one corner located at the western extreme of Ireland. How accurate are these triangles? Placing precise points on northern and eastern landmass extremes, the western corner of each triangle was then plotted on the western extreme of Ireland. All five western corner points are clustered on the Blasket Islands off the Dingle Peninsula, as shown in figure 6.7.

The repetition of isosceles triangle geometry between extreme points in the United Kingdom places the landmass in an unfamiliar light. With a modern perspective the geometry between landmass extremes appears accidental, but would the ancient perspective have been the same?

When Caesar arrived on English shores, the Druids told him that the land he was about to pillage was "triangular." In the turbulent times that followed, this understanding was erased forever; now, its origins can never be known. But the Neolithic circles are ordered in such a way that they repeatedly define geometric points located on these landmass extreme points, and a sequence of isosceles triangles result from joining them together.

The Neolithic method involves aligning with natural extreme points that appear to be geometrically ordered in their own right. Appreciation of this developed slowly. The following extract was written at the early stages of this research, long before this common method was fully recognized.

THE GREY WETHERS, DARTMOOR, ENGLAND

Dartmoor is the highest region in southwest England and higher than anywhere in England south of the Peak District. It is a wild, almost treeless terrain, a small wilderness of moorland and craggy rock summits called tors. For several thousand years the Grey Wethers double stone circle has marked a specific point in this wild landscape.

The highest summit on Dartmoor, High Willhays, can be reached by following a line bearing 314.15 degrees from the Grey Wethers stone circles. The total distance on the line is ten thousand megalithic yards, but the line passes over a second high point, Whitehorse Hill, which is found at the phi interval on the line.

Evidence of a broader geographic signature is also found here because the line bearing 314.15 degrees from the Grey Wethers continues over High Willhays to the Hartland Peninsula. This is the large coastal bump with a distinctly chevron-shaped profile on the north coast of Devon. The line does not reach the precise peninsula tip; nevertheless an isosceles triangle can be drawn between the tip of the Hartland Peninsula, the Grey Wethers stone circles, and the highest point on Bodmin Moor, Brown Willy.

The location of the circle leads directly to the recognition of the following natural topographical geometry in the local landscape: A bearing of 161.8 degrees can be taken from the base of Morte Point to the Grey Wethers. A bearing of 161.8 degrees can be taken from the tip of Morte Point to the tip of Prawle Point. From the base of Morte Point to the tip of Prawle Point is 61.8 nautical miles. A distance of 31.42 nautical miles can be measured from Morte Point to High Willhays. (These are tip and base measurements.)

The Langstone circle is on a line joining the Grey Wethers and the Fernworthy circle which is 6.18 nautical miles distant, give or take forty meters. The line runs tangential to each circle with great accuracy and measures out one nautical mile between the Grey Wethers and the Fernworthy circles, also with great accuracy.

And so it went on from one circle to the next across the moor, each time with alignments highlighting topographical extreme points, and when the distances and bearings between these points were measured in degrees and nautical miles, pi and phi digits resulted in profusion. These measurements may require one hundred meters tolerance, but they often defined both the tip and a base or flank point on a coastal peninsula, and again this occurred with such accuracy and frequency when measuring from a monument that it became clear the builders somehow recognized these as geometric points. Only gradually did the reasoning behind this emerge, and by this time we were studying only the largest monuments and the trail had taken us to the largest stone circle on the Iberian Peninsula, at Almendres in Portugal.

7

ALMENDRES

Among the remote parched hills of central Portugal stands one of the most striking and oldest examples of a double stone circle in the world (figure 7.1). It is the largest stone circle on the Iberian Peninsula, with more than ninety surviving stones arranged in two ellipses that, according to the notice board at the site, were erected between the fifth and fourth millennium BCE (figure 7.2). Even with such an imprecise date, this is perhaps the oldest stone circle in Europe.

The leaves on the cork and olive trees rustle in the breeze around Almendres, which is situated in a region of relatively arid hills, some of which rise to more than four hundred meters. But the circles of Almendres, although on a ridge of high ground that affords wonderful views, are not located at the top of the hill. At first glance the site

Figure 7.1. The double stone circle at Almendres. Photo by Jlrsousa.

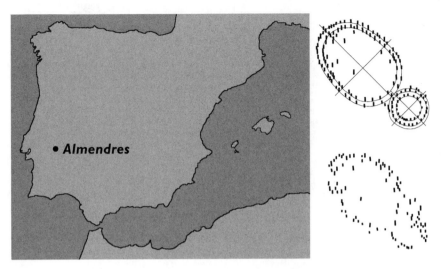

Figure 7.2. The location of Almendres and
schematics of the stone circle.

appears to have been randomly chosen, almost as if the builders couldn't
have been bothered to climb the few hundred meters to the highest
point. But when the location of Almendres is considered in relation to
pi- and phi-digit measurements, a truly remarkable picture emerges. It
appears very clear from the precise position of the monument that it
was located by people who were completely conversant with geography
and who had diligently searched for a location from which pi and phi
digits would result when measuring to extreme topographical points.

To use a musical analogy, it is as if a theme tune had been written
by an ancient master composer and the builders at Avebury, Rollright,
Stanton Drew, Brodgar, Loughcrew, and Stonehenge had all heard the
refrain, had the tune in their brain, and had reproduced it to the best of
their considerable ability. The builders at Almendres also knew this tune.

MEASURING ALMENDRES

The north coast of the Iberian Peninsula is a relatively straight line
running roughly east/west, but the line is interrupted by a slight bulge
where the northern extremity of the Spanish coast is found. On this

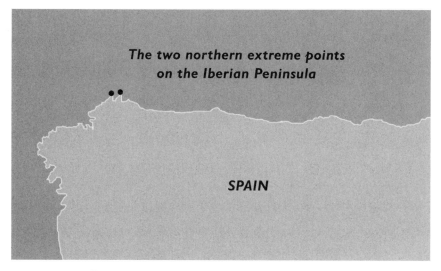

Figure 7.3. The two pointed peninsulas at the
northern extreme of the Iberian Peninsula.

bulge there are two peninsulas, about fifteen kilometers apart. These
peninsulas are the two northernmost extreme points on the Iberian
landmass; they are marked with dots in figure 7.3.

In light of the Neolithic measurements discussed so far, it seems that
even the most imaginative combination of chance, accident, serendipity,
coincidence—call it what you will—fails to account for the following:

1. The most northerly point on the northern peninsula is 314.2
   nautical miles from Almendres.
2. A bearing of 3.142 degrees can be taken from Almendres to this
   peninsula.
3. The second northerly peninsula can be reached with a bearing of
   1.618 degrees.*

These measurements surely confirm a common code of practice in
Neolithic Europe. There is only one place Almendres could be located
in order to achieve this distinction, and that is exactly where it is. It

*Measurement taken to the eastern limit of the peninsula point.

would seem that, far from being lazy about their choice of site, the builders were being meticulously accurate. Looked at in detail the bearing and distance with pi digits serve to locate the tip and a base point of the northern peninsula.

It is also possible to measure 314.2 nautical miles from Almendres to the highest mountain summit in northern Spain, Torre de Cerredo. The result is an enormous isosceles triangle drawn between this highest northern point, Almendres, and the most extreme northern point on the Iberian Peninsula. This triangle has the same leg measurements as those used between Rollright, Ben Nevis, and Ben Macdui. To complete the geometric motif recognized in England, this isosceles triangle should be attached to a right triangle. The right angle is found by joining Almendres to the base point on the peninsula (shown in figure 7.4) and the rocks on the tip of the second of the northern peninsulas.

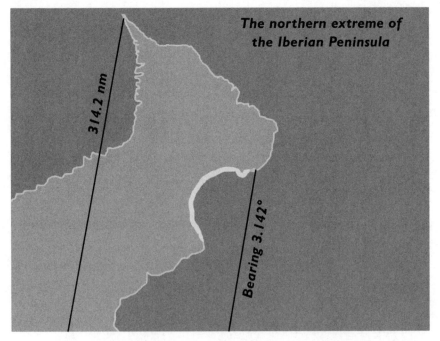

Figure 7.4. The line measuring 314.2 nautical miles from Almendres to the peninsula tip, and the line bearing 3.142 degrees to the base point on the peninsula; the distance between the two points is about 2.5 kilometers.

Neolithic people appear to have been demonstrating their ability to define the landmass extremes we see today in the context of an underlying geometric pattern related to high points. Three methods were used to confirm the measurements from Almendres: Google Earth, the Vincenty formula, and the Haversine formula (which measures straight line distances—along great circles—on a true sphere). All three concur within one hundred meters; with this small tolerance the distance from Almendres to the farthest rock on the northern peninsula is 314.2 nautical miles. Modern technology forces a revision of past ability, but this essential revision is resisted at every turn by mainstream academics with the turgid refrain, "It must be a coincidence."

When he measured the Neolithic stones of Britain and Brittany fifty years ago, Thom realized that he was studying the achievements of an extraordinarily scientifically intelligent, well-informed, and imaginative people. He found this by measuring. His insight led him to the recognition of their "superior" brainpower, as he put it. There was a little flurry of interest before he was all but completely ignored. Many among the following generation of archaeologists have continued to assume that all the thousands of tons of stone and earth that were dragged hither and thither in Neolithic Europe were serving the purposes of primitive rituals. This may be the case, but somebody at the time needed a persuasive reason for locating these extremely heavy "ritual" stones at a specific place. Archaeologists have found no common denominator to explain why the remote and seemingly arbitrary locations for stone circles were chosen. One coherent explanation is found by recognizing the brainpower of these people, just as Thom did. There is further confirmation of this intellect at Almendres.

A measurement of 161.8 nautical miles can be taken from Almendres to the highest mountain in central Spain, Almanzor. This line extends to the base, or flank, of the massif.

Allowing that Almendres was a location chosen by people with a clear understanding of their whereabouts on the planet, a broader topographical picture emerges. The highest summit on the Iberian Peninsula is Mulhacen in the south, but there are no direct alignments, or pi- and phi-digit measurements, from Almendres to this summit. However, if the now-familiar process of joining the specified points independent of the monument is followed, once again natural topographical geometry is revealed.

If the two peninsula points on the north coast are joined to Mulhacen's summit, both lines pass over the flanks of Almanzor. Thus a natural alignment of topographical extremes is once again discovered through this Neolithic method, as shown by the lines in figure 7.5.

As with the British monuments, the result of following this process at Almendres is to discover natural geometry; natural topographical extreme points are in alignment. If these are indeed a few echoes of the Neolithic theme tune, the builders at Almendres have once again found an extraordinary focal point in the landscape that draws attention to the ordering of topographical extremes in nature.

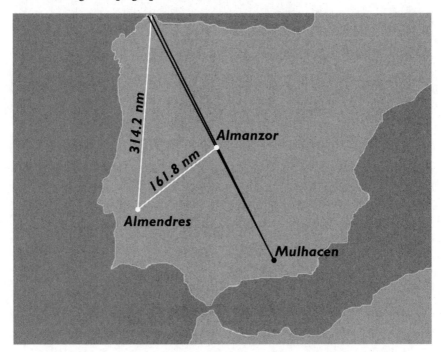

Figure 7.5. Two lines begin on the two northern extremes of the landmass and pass over the flanks of Almanzor en route to the summit of Mulhacen. The distance from the specified points to Almendres is also shown.

CABO DE ROCA AND DANTE ISLAND

Almendres is about 127 kilometers from the western extremity of the Eurasian landmass, Cabo de Roca in Portugal. A line from this point passing over Almendres extends to the eastern landmass limit of Africa, which, although it is a peninsula attached to the mainland, is called Dante Island. Consequently, Almendres is aligned between two cardinal continental landmass extremes, ^-^-^, but not absolutely precisely. The line that passes directly over Almendres extends from the base of one peninsula to the tip of the other. The line is over 6,500 kilometers in length, and it joins the two cardinal continental extreme points from tip to base, as shown in figure 7.6.

The two continental coastal extremes are specified as geometric points by this alignment process, just as the two northern extremes of the Iberian landmass were specified by pi- and phi-digit measurements. In this case the process of joining the specified points independently once again produces information about geometric order in the natural world. The secondary northern Iberian peninsula was found on a bearing of 1.618 degrees from Almendres, and if this extreme point is joined to the two continental extreme points, at Cabo de Roca and Dante Island, the result is a vast intercontinental right triangle accurate to about 250 meters on the ground when measured on Google Earth.

So once again the extreme points initially specified by a monument are joined independently, and geometry is revealed between these natural topographical extreme points on the Earth. Impossible though it may seem, it becomes entirely reasonable to recognize that Almendres was located by people aware of these measurements. And if disbelief is suspended, the motive behind their actions also becomes clearer. At every monument visited, the highlighted extreme points have produced unusual geometry when they are joined independently. While no explanation of this Neolithic ability can be offered, the same process has led to the same conclusion in every case. As Thom discovered, this Neolithic practice involved highly accurate measurement: each monument location identifies topographical geometry unrecognized in the

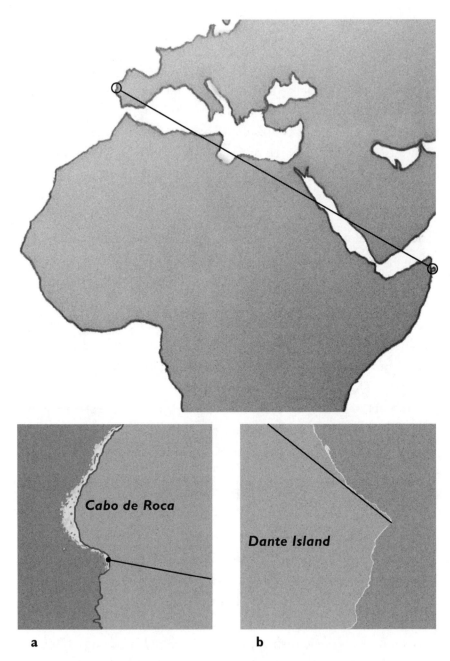

Figure 7.6. These images depict the western landmass extreme of Eurasia at Cabo de Roca (a) and the eastern landmass extreme of Africa on the eastern tip of Dante Island (b). The Almendres circle is located on a line joining the two continental extreme points together.

present day. The key is the monument builders' allegiance to topographical extremes. The degree of accuracy is beyond perplexing, not least because these are present-day alignments. There is no known method by which this could have been achieved at the time, but the weight of evidence demonstrating that it was achieved is considerable. Each monument ultimately provides an insight into the repetitive geometric patterns that are currently created between the highest and most extreme points on a landmass. This is written down as clearly as any document or manuscript from the past, but the text is translated by defining the lines and then measuring between the ^ symbols on them.

Almendres specified the mountains of Torre de Cerredo and Almanzor with pi- and phi-digit measurements. These mountains are specified again by another stone circle, but this time the two mountains are in direct alignment with the monument. The Neolithic circle is located at Msoura in Morocco; it is the largest known stone circle in the country.*

MSOURA

In the northern region of Morocco, about 40 kilometers to the south of Tangiers, the ancient stone circle at Msoura (or Mezorah) is more than fifty meters across. The alignment of these stones with the summits of Torre de Cerredo and Almanzor is accurate to a few hundred meters on the ground over a distance of 870 kilometers. A survey by James Watt Mavor, Jr., in the 1970s established that the shape created by the Moroccan stones is an ellipse based on a 37-35-12 Pythagorean triangle.[1] Thom had already noted a number of monuments in Britain created by using this triangle.

The two-summit alignment with Msoura finds a point on the flank of Almanzor at the phi interval on the line. This line also forms one side of a right triangle drawn between Msoura, Torre de Cerredo, and Aneto, the highest mountain in the Pyrenees. But the third geometric

---

*Msoura may be the only stone circle in Morocco; there are others in Libya and Egypt.

point is not on the summit of Aneto; it is found in the valley at the base of this precipitous peak. Apparently Msoura's builders also knew the Neolithic signature tune and adopted the same tip-to-base leitmotif.

The distance from the Msoura stone circle to the summit of the highest point on the Iberian Peninsula, Mulhacen, is 161.5 nautical miles, about five hundred meters short of 161.8 nautical miles.

The Moroccan circle carries many of the familiar keynotes: the alignment of regional high points with the monument, the phi interval on the line, a phi-digit measurement to the highest mountain, and the directive to natural landscape geometry in the form of the right triangle between high points that includes a tip-to-base measurement. From virtually any modern perspective, it is extremely difficult to imagine that measurements of this scope and accuracy were possible in prehistoric Africa. Yet, equally, is it reasonable to assume that this replication results by chance?

Surely the largest Neolithic site in Europe, at Carnac in Brittany, should supply a convincing answer. If we have indeed caught echoes of a Neolithic refrain, all the keynotes can be anticipated at this, the most extraordinary of all the European Neolithic monuments. And there the overland lines extending to high points do not need to be imagined; the builders have set them down in stone.

At Carnac the Neolithic theme tune can be measured in decibels.

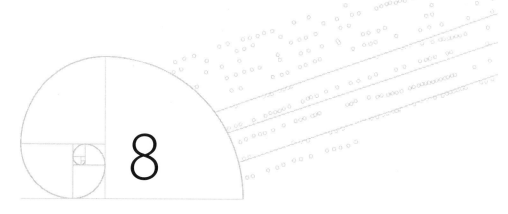

8

A LINE TO EVEREST

CARNAC REVISITED

Neolithic people habitually transported large stones over several kilo-meters and stood them upright. At Carnac more than three thousand were set in a series of lines spanning more than three kilometers. Le Menec West, as shown in figure 8.1, runs "straight" for about five hundred meters. Close by, at Locmariaquer, they excelled themselves. The Grand Menhir (great standing stone), though now broken and recumbent, possibly as the result of an earthquake, is a granite monster eighteen meters long and weighing some 280 tons. As this particular type of granite is unknown on the Locmariaquer Peninsula, experts have concluded that the monolith must have been shifted several kilometers to reach its destination.

RITUAL BELIEFS

These "primitive" people are understood to have been in the earliest stages of peasant living; stone, wood, and bone were their principal materials, pots rendered from baked sea mud were their luxury culinary items, and they accessorized with stone beads and bangles. If life was short and mere survival a struggle, what could have driven them

Figure 8.1. Le Menec at Carnac, where we started.
Photo by Jos Le Doaré.

to waste precious time moving rocks? Tens of thousands of hours of labor were dedicated to creating these lines at Carnac. Today, their purpose is variously suggested to have been "ceremonial," "processional," "devotional," "astronomical," or "funereal," yet none of these fully explain the effort. A bemused tourist, having absorbed all such interpretations offered in the local museum, wrote in the visitors' book, "We can fathom neither their strategy nor their reasoning. Neolithic man has checkmated us!"

Is it credible that ceremonies, rituals, or festivals provide an explanation for standing thousands of stones upright in the ground? Surely an animal-skin marquee would have sufficed? What degree of obsession ate into the heart of these Neolithic people and so overwhelmed them, generation after generation, with the need to shift vast rocks into rings, rows, dolmens, and monoliths or to pile up great mounds and earth banks? Nobody has ever satisfactorily explained these things. Was Carnac really "a kind of temple," as some archaeological literature suggests,* and if so, what on Earth were these people worshipping? The present-day interpretation has determined that these monuments were created with cultish rituals of varying kinds in mind. It is a best guess and it may be true, but it is not what the stones explicitly communicate.

THE ENIGMA OF CARNAC
Following the Lines

The Carnac stones are set in lines that stretch as far as the eye can see—and farther. Like an army battling with time itself, rank upon rank of these stones are merely the survivors in a project of vast proportions; the lines range over the low uneven land, traveling up and down slopes, over a shallow ravine, and out into the distance. It is obsessive line making, not in any abstract way, but physically, on the ground itself. How else could the Neolithic people describe geometry in the landscape more clearly?

---

*This quote is from the free pamphlet distributed at the Musée de Préhistoire in the center of Carnac.

The dramatic fault line between orthodox and alternative theories about Carnac could not be clearer. To the archaeologist Aubrey Burl, these lines are "haphazard . . . prehistoric accretions whose successive builders were not obsessed with exactness."[1] In contrast, the engineer Alexander Thom "cannot help being impressed with the great accuracy with which the lines were laid out."[2]

Thom supervised the most comprehensive survey of the stones ever undertaken and, careful to be as exact as possible, admitted he was hampered by the fact that an estimated 80 percent of them have been re-erected. He also found that modern construction "has damaged the alignments badly . . . moving some of the stones without leaving any record of their original positions." He was further hindered by "vandalism" and also asked, "Where the stones had all fallen one way, is it possible that the wrong end was lifted so that the whole line was displaced?"[3]

Although the stone lines meander considerably, there is a clear visible correlation between the ancient stone rows and lines extending to the world's three highest mountain summits: Mount Everest, K2, and Mount Kanchenjunga (see figure 8.2 and chapter 1). But the evidence that these lines were purposefully directed at these mountains does not rest on the visible correlation. The common Neolithic code appears clearly through measurement on Google Earth. The following figures are rounded to one decimal place.

1. The bearing from the summit of Mount Everest to Le Menec is 314.0 degrees.
2. The bearing from the summit of Mount Kanchenjunga to Le Menec is 314.4 degrees.
3. The mean bearing from these two mountains to Le Menec is therefore 314.2 degrees.

The code of practice found among the Neolithic circle builders elsewhere in Europe echoes resoundingly at Carnac with these pi-digit measurements; and the lines are still visible on the ground. There they

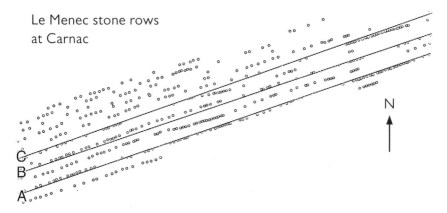

Le Menec stone rows
at Carnac

Figure 8.2. The stone lines at the beginning of Le Menec.
Lines A–C indicate the direction to the three highest summits in the
world. This diagram of the lines also appeared as figure 1.2.

remain, witness to an unseen aspect of our ancient history. Carnac is a location chosen by people who knew exactly where they were. The monument introduces an amazing global pattern, an insight into the foundations supporting the ancient worldview of an ordered Earth. It communicates this view through measurement.

To the east lie the Himalayas; what lies to the west?

AMERICAN PEAKS

Here are the measurements from Le Menec to the highest mountain summits on the North and South American continents:

1. North America: The distance from Le Menec to Mount McKinley's summit is 4,002 nautical miles.*
2. South America: The distance from Le Menec to Mount Aconcagua's summit is 6,000 nautical miles.

---

*At the time of writing, President Obama proposed to rename Mount McKinley, giving it back the original local name Denali.

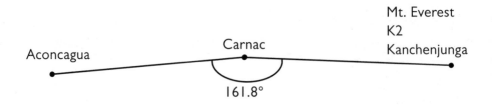

Figure 8.3. The 161.8 degree angle at Carnac.

Thom noted that Neolithic builders were sticklers for whole number measurements. The highest peak in the Americas, Mount Aconcagua, is signified by such a measurement, just as the highest mountains in Asia are signified by alignment. But the remarkably calculated nature of this Neolithic work is found by making a further measurement. When the lines to the American and Asian summits are drawn from Carnac, the angle between them is 161.8 degrees (see figure 8.3).

The phi-digit angle in figure 8.3 is created by two lines starting at Le Menec stone circle. The first line is six thousand nautical miles long and extends to the summit of Mount Aconcagua. The second line follows the stone rows and has the mean bearing of lines extended to the three highest summits in the world. The angle between the lines is 161.8 degrees, rounded to one decimal place.*

The scale of Carnac reflects its global reach. It is big because it needs to be. Carnac has identified the highest summits in Asia by alignment and pi-digit bearings and the highest summit in the Americas, Aconcagua, with a phi-digit angle and six-thousand-nautical-mile distance. What happens on a third continent? Is the highest summit in Africa similarly identified?

---

*The bearing given on Google Earth from Le Menec stone circle to Aconcagua's summit is 231.89 degrees. The mean bearing from the same point to Asia's three highest summits is 70.05 degrees. The difference is 161.8 degrees, rounded to one decimal place.

AFRICAN PEAKS

The snows of Kilimanjaro are thawing and may soon be consigned to memory, living on only in a few ice core samples in the freezer at Ohio State University and in the celluloid annals of Hollywood. In the eponymous short story by Ernest Hemingway, a writer dying slowly from an infected leg wound lies in the shadow of Africa's highest mountain and reflects on his life as vultures circle patiently overhead. Today the vultures remain, but the snows are melting fast. There are various estimates of the glacial recession on Kilimanjaro and a reasonable amount of agreement that it has lost about 75 percent of its ice cover since 1912.

A line from the summit of Mount Kilimanjaro extended to Le Menec stone circle continues over the Atlantic Ocean to the southern extreme of Greenland. The line arrives at Cape Farewell, the most southerly peninsula in Greenland, on Egger Island. This line is nearly ten thousand kilometers in length, and it passes within two kilometers of Africa's northern extreme point at Cape Angela.* If the line is redrawn from the summit of Mount Kilimanjaro, this time passing over the tip of Cape Angela, it also continues to the southern peninsula on Egger Island and passes over the coastal peninsula at Carnac. The alignment of Le Menec with Mount Kilimanjaro and Egger island therefore duplicates the natural alignment of Cape Angela with the same two extreme points. The southern extreme of the contiguous landmass of Greenland is about fifty kilometers to the west of Egger island; measured on Google Earth it is 1,618.6 nautical miles from Le Menec stone circle to this landmass extreme point.

---

*The name of Africa's northern extreme point is subject to change from one authority to the next. It was and is known as Ras Ben Sakka, Ras al-Abyad, and Cape Engela, as well as Cape Angela.

THE GEOMETRY OF GRAND DESIGN

The highest mountains on the three largest continents in the world have now been specified from Carnac, but to what purpose? If our understanding of this Neolithic process is correct, then these three great mountains of the world, Mount Everest, Kilimanjaro, and Aconcagua, should be treated as specified geometric points and joined by lines. This process is an integral part of the code of practice, and if past examples are anything to go by, it should reveal evidence of order, or pattern, on the Earth's surface.

And it does.

CONTINENTAL SUMMITS

A distance of 6,180 nautical miles can be measured from the summit crater of Mount Kilimanjaro to the summit of Mount Aconcagua.

The Neolithic code reveals an extraordinary phi-digit measurement on the Earth's surface. This process continues to illustrate aspects of the Earth's geography from a distinctly different perspective. It is instructive; a phi-digit measurement separates these two great continental summits. The measurement further attests to the extraordinary global awareness evidently possessed by the people who found the location at Carnac.

To continue the process of joining the specified mountains together, a line is drawn from Mount Aconcagua to Mount Everest. This line passes over the largest freestanding volcano in the world, Mount Elgon.[4] Thus, once again a geometric order, ^-^-^, is found by joining the specified points. And again this process yields a phi-digit measurement; a distance of 6,180 nautical miles can be measured from the center of Mount Elgon's crater to the summit of Mount Aconcagua.

The result of following this simple code is to discover an enormous isosceles triangle with legs measuring 6,180 nautical miles that can be drawn around the globe between the summits of Mount Aconcagua, Mount Kilimanjaro, and Mount Elgon. And when shifted to the base

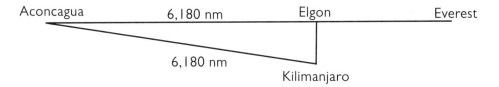

Figure 8.4. The isosceles triangle joining Mount Aconcagua, Mount Kilimanjaro, and Mount Elgon, with one leg extending to Mount Everest.

of Mount Elgon, one leg of this triangle extends to Mount Everest. This surprising global geometry between continental high points is illustrated in figure 8.4.

So there is a join-the-dots form of communication illustrating the unusual geometric relationship that exists between the highest summits in Asia, Africa, and the Americas. And each of these summits is specified by alignments or pi- and phi-digit measurements from Carnac. How different would the modern mind-set be if geography was illustrated in this manner at school?

Mount Elgon is identified by the Carnac geometry, and so the two locations are joined with a line. This final straight line creates one side of another enormous isosceles triangle, this time between Mount Everest, Carnac, and the base point already established on Mount Elgon.

GREAT VISIONS

A great scientific instrument lies sprawled over the entire surface of the globe. At some period, thousands of years ago, almost every corner of the world was visited by people with a particular task to accomplish. . . . The key to the mystery must surely lie in the study of the great pattern itself and its relation to the subtle forces of the landscape.[5]

These thoughts from the late John Michell are the conclusions of his lifelong quest to understand the motivation of monument builders around the world. His studies led him to recognize the global reach of these builders from prehistory. In consequence he says:

> We are suddenly faced with an unavoidable truth. Our current orthodox model of history is shown to be plainly and absurdly wrong, so completely wrong as to throw into disarray our present beliefs concerning the capabilities of prehistoric man.[6]

John Michell, like Alexander Thom and many others, was largely ignored.

It cannot be denied that the builders at Carnac were concerned with creating long and straight lines over the landscape. Measuring straight lines from Carnac makes it increasingly difficult to imagine how this extraordinary location could have been chosen in ignorance of its qualities and how this ignorance could extend to creating lines of stones directed to Mount Everest. The brainpower that gave rise to these astonishing structures is discernible through measurement. Measured geometry was the Neolithic mother tongue. Thom wrote, "They were intensely interested in measurements and attained a proficiency which as we shall see is only equalled today by a trained surveyor. . . . They concentrated on geometrical figures which had as many dimensions as possible arranged to be integral multiples of their units of length."[7] And lacking pen and paper, they voiced their passion through their monuments. The ancient world is still alive with this topographical language: circles, mounds, and pyramids are geometric points identifying a geometric worldview.

WORLDVIEWS

The measured order on the veins of a leaf or the scales of a fish comes as little surprise, yet the conceptual leap toward recognizing something similar on the surface of the Earth is clearly enormous. But the evidence from early recorded history is also clear on this point: there was no great conceptual leap for these people. From the dawn of history, it was understood that the planet was "ordered." Be it leaf, fish, or landscape, nature was as one. The genius of the monument builders was then to discover a location that illustrated, or demonstrated, the true nature of the ordered Earth. The natural pattern is found through alignment and through sexagesimal-based measurement, and at its foundation are the digits 3142 and 1618 (and/or 618). Carnac is at a focal point from which measurements to topographical extremes *repeatedly* produce alignments and pi- and phi-digit measurements.

The principle of aligning a monument with regional high points and with extreme coastal points was identified in Britain. It is replicated at Carnac for a second time with an alignment to Africa's southern extreme, Cape Agulhas.

The line from the Le Menec stone circle to Cape Agulhas passes over Mount Tahat, the highest summit in the western Sahara. The two extremes are thus specified by their ^-^-^ alignment with Carnac, and the distances on the line are consequently measured. The distance from Le Menec to the coastal rocks at the southern extreme of the landmass of Africa is $1.618 \times 3{,}142$ nautical miles, accurate to two hundred meters on the ground.

From Brittany the monument reaches out over the continent of Africa to its southern point. In turn a line from Carnac passing over Africa's northern extreme, at Cape Angela, continues to the flank of Mount Kilimanjaro. The continent's northern and southern coastal extremes are specified through alignments with Mount Kilimanjaro and Mount Tahat. In this way we are invited to join the two specified coastal extreme points.

THE AFRICA LINE

So the northern and southern coastal extremes of the African landmass are joined together; a straight line is drawn from the tip of Cape Agulhas in the south to the tip of Cape Angela in the north. When this line is extended, it goes to a third continental extreme point, the northern extreme of the American continent at Cape Columbia (figure 8.5).

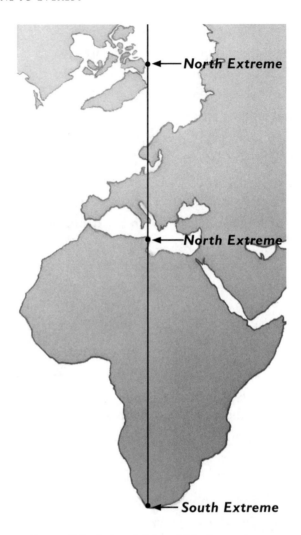

Figure 8.5. A line joining Africa's northern
and southern extremes extends to America's
northern extreme peninsula, Cape Columbia.

Therefore, by following this process three continental cardinal extremes are found sharing a single straight line: ^-^-^. The code
of practice used to identify this alignment is identical to the process
adopted at all the other, smaller Neolithic sites, but at Carnac the canvas
is broader. This intercontinental line is just the beginning of a process
that identifies an ordered distribution of continents on the Earth.

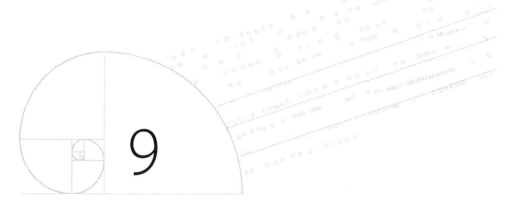

9

INTERCONTINENTAL

Throughout history the limits of human endurance and endeavor have been tested at the limits of the Earth itself. The dramatic accounts of the first men to venture to the poles of the Earth or to the Earth's highest summits attest to this. In many cases the extremes of the continents have proved equally testing, perhaps none more so than Cape Columbia at the northern limit of North America.

GEORGE WARD HUNT

Just off the coast of Cape Columbia, Ward Hunt Island lies embedded in the Arctic ice. This frozen lump of rock is the northernmost of all North America's small islands, a dot in the ice less than seven kilometers from east to west and four kilometers wide. It shares the latitude of Cape Columbia at the northern tip of its much larger neighbor Ellesmere Island.

In 1876 Sir George Strong Nares, leader of the British Arctic Expedition, wintered his ships the *Alert* and the *Discovery* in two bays off Ellesmere Island. He sent out sledging parties to the east and west and put Albert Hastings Markham in charge of a northern party endeavoring to reach the North Pole. Markham's men pushed forward in appalling conditions, shoveling and ice-picking their way across the treacherous

Figure 9.1. The *Alert,* one of two ships in the 1876 British
Arctic Expedition lead by Sir George Strong Nares.
Photo taken by Nares near Cape Beechy in 1878.

frozen wastes to reach 83 degrees north—a first—before falling prey to
scurvy and frostbite. A rescue party reached the team in the nick of time.

Pelham Aldrich, the leader of the western sledge party, also had
to be rescued, but he at least had the consolation of being the first to
reach Cape Columbia, now recognized as the northernmost extreme of
the North American continent. Following these ordeals Nares realized
that his men would not survive another Arctic winter. He set course for
home, blasting his way through the ice with gunpowder.

Nares's men were the first Westerners to sight Ward Hunt Island in
the frozen sea off Cape Columbia. They named it as a tribute to George
Ward Hunt, who was First Lord of the Admiralty at the time. Ward Hunt
later became Chancellor of the Exchequer, and he gave the British people
a quirky legacy. Prior to delivering his first (and last) Budget speech, he
arrived at the House of Commons only to find he had unwittingly forgotten his briefcase containing the essential papers. This unfortunate
lapse gave rise to a still-current tradition: on Budget Day the Chancellor
will raise his briefcase to shoulder height upon leaving 11 Downing Street

to assure the public that the incumbent Chancellor's memory is more reliable than that of the unfortunate George Ward Hunt.

INTERCONTINENTAL LINES

The line that joins the southern extreme of Africa to its northern extreme continues around the Earth to Cape Columbia; thus, three continental extremes are found in alignment. But if the line is extended, it passes over Cape Columbia and continues over the icy wastes of the Arctic Ocean, into Alaska, and then onward directly to Mount McKinley, passing over the East Buttress of the mountain two nautical miles from the summit. This line therefore joins North America's highest mountain with three peninsulas, and each peninsula is located at the cardinal limit of a continent. A natural intercontinental ^-^-^-^ has been identified by joining the two cardinal extremes of Africa both of which are specified by alignments from Carnac. On Google Earth the end of this line (extending all the way from Cape Agulhas) can be transferred to the summit of Mount McKinley, and it then passes over both Cape Columbia and Ward Hunt Island. It was extraordinary to find that using Google Earth's outline of Ward Hunt Island, it is possible to measure from the island's eastern tip for exactly 3,142.02 nautical miles to the northern limit of the landmass of Africa, at the tip of Cape Angela.

THREE PHASES

The result of measuring between extreme points is informative and intriguing. The procedure leading to the recognition of these odd measurements provides a striking illustration of how the code of practice operates in three simple phases.

> Phase One: Lines joining Carnac with dominant mountain summits pass directly over the northern and southern extremes of Africa. These two continental extremes are thus highlighted, or specified, by this alignment process.

Phase Two: Having used a Neolithic monument—in this case Carnac—to find these two specified extreme points in Africa, they are then treated independently. The points are joined with a line, and the line is extended. In this case alignments with three further topographical extremes are found, Cape Columbia, Ward Hunt Island, and Mount McKinley. In this way the Neolithic alignments reveal a further natural line with four continental extreme points located on it; thus the repeated motif of: ^-^-^, and in this case with one more, -^.

Phase Three: In phase three the lines are measured using bearings and nautical miles, and the results yield pi or phi digits with significant frequency.

So these digits can now be anticipated when measuring the intercontinental line specified by the Carnac alignments. And as we have seen, the distance between Cape Angela and Ward Hunt Island is 3,142 nautical miles.

A SIMPLE CODE

Stripped to its bare essentials, it is a simple but intelligent code.

1. Three-point alignments with the monument are identified: ^-^-^.
2. Aligned points are joined independently and measured.
3. Phi and pi digits result with significant frequency.

A reader tracking back through the earlier illustrations from Britain (supplemented in the appendices) will find that the same code gradually becomes apparent from one stone circle to the next. On each occasion measurements on the lines between the specified points yield a frequency of pi and phi digits that cannot readily be attributed to chance.

The three phases of the process, repeated at each site, provide clear evidence of a common practice adopted by these Neolithic builders, and a language of symbols is created by following it. The monument can then be removed from the picture, and the specified points in the landscape produce geometry with pi- and phi-digit measurements independently. The language therefore communicates information about an apparently ordered arrangement of these natural topographical extremes. Aconcagua to Kilimanjaro is 6,180 nautical miles, and Cape Angela to Ward Hunt Island is 3,142 nautical miles, and so on. The continental topographical extremes may appear, if you have that frame of mind, to be ordered by their geometric arrangement and fashioned by the repetitive pi- and phi-digit measurements between them.

The Neolithic people appear to have questioned the true nature of Nature herself, and when measuring along these lines, the Earth's surface is found to be intriguingly married to measured geometry. They developed a simple and sensitive way of throwing light on these features of nature and of preserving their awareness for posterity. Stones endure.

A TALE OF THREE WILLIAMS

In 1868 the U.S. Secretary of State, William Seward, gave Russian diplomat Edouard de Stoeckl a check for 7.2 million dollars. For his money he got 663,268 square miles of land on the northwestern extreme of the North American landmass, at a value of approximately two cents an acre, a price many of his detractors considered extortionate. Seward's Folly remains today one of the nicknames for Alaska, which in 1959 became the forty-ninth state in the union.

To his end Seward remained convinced that the acquisition of Alaska for the United States was one of his greatest achievements. He died in 1872, just as events were beginning to justify his conviction: gold was discovered near Sitka in southeast Alaska, sparking the Cook Inlet gold rush.

In 1896 a prospector was panning for gold in the Susitna River valley between the Talkeetna Mountains and the Alaskan Range. His name was William Dickey. Pausing from his work he realized he was looking at a truly big mountain known to the local Athabascan tribe as Denali, "the high one." He renamed it Mount McKinley, in tribute to his champion and a proponent of the gold standard, the Republican William McKinley, soon to be elected U.S. president.

Mount McKinley is not an easy mountain to get to know. Twin-peaked, the southern summit elevation is 20,320 feet (6,194 meters), the highest point on the North American continent. In 1910 four local men, known collectively as the Sourdoughs, claimed the first ascent of the north peak. They carried with them a fourteen-foot spruce pole, and for the final assault were fortified with a bag of doughnuts and hot chocolate. Their claim was not verified until Hudson Stuck's ascent of the south peak in 1913, when, through his binoculars, he reportedly saw the spruce pole still in place.

A LINE TO MCKINLEY

Having found the intercontinental line between the extremes of Africa and North America, the remainder of the line, from Ward Hunt Island to Mount McKinley, was also measured. A distance of 1,571 nautical miles (i.e., one-half of 3,142) can be measured from Ward Hunt Island to Pioneer Ridge on Mount McKinley's north flank. The intercontinental line is illustrated in figure 9.2.

Applying the ancient code of practice has led to the discovery of this intercontinental line and to the strange measured intervals on it.* It is geography, but not as we know it.

---

*The tolerance on the intercontinental line is 0.07 degrees bearing from Cape Agulhas. This, in close-up, gives two lines. The first line, bearing 351.48 degrees from Cape Agulhas, passes over the tip of Cape Angela, continues to Cape Columbia, and finishes on the flank of Mount McKinley. The second line, bearing 351.55 degrees, passes over Cape Columbia, over Ward Hunt Island, and on to the summit of Mount McKinley. This second line passes over and aligns with an unusual fingerlike promontory at Africa's northern extreme, about eight kilometers from the tip of Cape Angela.

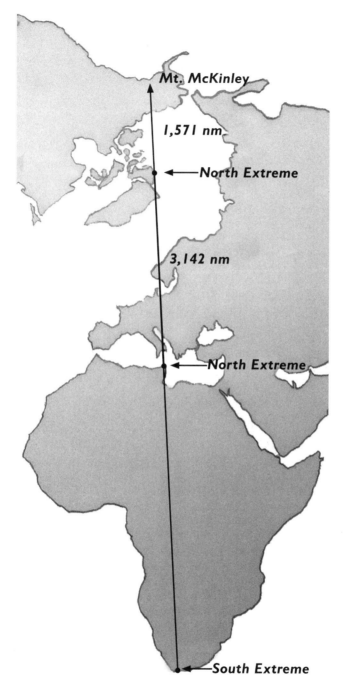

Figure 9.2. The straight line joining four
continental topographical extreme points.

FOUR CORNERS

It seems that the chief decision dictating the choice of a monument's location was reached with a view to revealing the topography of the globe as something with measurement at its heart, that is, a designed or ordered surface.

Among the earliest written accounts of the Earth, a reference to the "four corners" is found repeatedly. It is in the Bible and in myths from Greece, Egypt, Mali, the New Hebrides, the Pacific Marshall Islands, Guatemala, China, Scandinavia, and at least twelve native North American tribes.* It is popularly assumed that the expression "the four corners of the Earth" refers to the cardinal points of the Earth, but not everyone thinks so. The expression appears twice in the Bible, and according to Hebrew scholar Joan Sloat Morton, Ph.D., the word *corners* is more correctly translated as "extremities."[1]

Clearly the Earth sphere does not have four corners, and how can anything spherical have four extremities? But the *landmasses* on the sphere all have these cardinal extremities, and with this understanding the phrase immediately makes sense. Each landmass has four cardinal extreme points, or corners. The ancient monuments repeatedly indicate these precise corner points by creating alignments with them, and when the corner points are joined with lines and the intervals on the lines are measured, the Earth begins to appear mysteriously ordered.

Is this really the case? Do the continental corners produce a pattern?

KOMSOMOLETS

Komsomolets is a truly extreme point: the last chunk of rock and ice between the northern tip of Asia's landmass and the North Pole. It was named in honor of members of the Komsomol, the Communist Union of Youth. The island appears to be enormous on a Mercator

---

*These include the Pawnee, Navajo, Lenape, Yuma, Winnebago, Pomo, Dieguenos, Yuki, Cahto, Apache, Jicarilla, and Omaha.

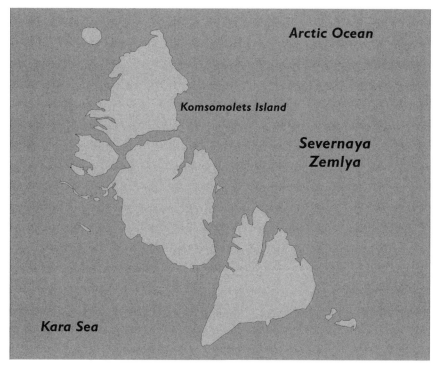

Figure 9.3. Komsomolets Island extends to the most northerly
point in the Severnaya Zemlya archipelago. These islands
are located directly to the north of Cape Chelyuskin, Asia's
northern landmass extreme. Komsomolets, 130 kilometers
from north to south, is the most northerly point.

map due to the distortion at high latitudes. It is actually about the
size of Majorca.

Despite—or perhaps because of—its extreme location, this region
has seen a fair number of dramas and tragedies. In 1922 twenty-nine-
year-old Soviet explorer and geologist Nikolay Urvantsev discovered
some papers strewn about on the shore of the Kara Sea, close to the
mouth of the Zeledeyeva River. (He believed they had possibly been
scattered by a curious bear.) It transpired that these papers contained
important scientific data; they were mail from Roald Amundsen's
Arctic expedition of 1918. Two men, Peter Tessem and Paul Knutsen,
had been entrusted with their safekeeping.

Left behind on Asia's northern contiguous landmass extreme, at Cape Chelyuskin, the two men had instructions to make their way by sledge some eight hundred miles southwest to Dikson, but this could only be achieved when the Kara Sea had frozen over. The journey should have taken them no longer than eight weeks, once conditions turned in their favor, but the two men were never seen alive again.

One frozen body was discovered in July 1922. Identified by a gold watch engraved with his name, it was assumed to be the corpse of Peter Tessem; he was buried close to where he was found in a grave marked with a cross of wood. In 1958 his remains were relocated and reburied under a granite monument at Cape Chelyuskin, with a dedication in both the Russian and Roman alphabets. If these are indeed the remains of Peter Tessem, one can only speculate as to the final resting place of Paul Knutsen. Perhaps he was swept to the icy wastes of Komsomolets off to the north. If so, he may rest at a very interesting point on the planet.

ORDERED POINTS

Komsomolets in Asia and Ward Hunt Island in North America are both distinguished for the same reason: their location. They are both the most northerly small island off the northern extreme of a continent.

It was surprising to find that the measurement of 3,142 nautical miles from Cape Angela to Ward Hunt Island is replicated by a measurement of 3,142 nautical miles from Cape Angela to Komsomolets. Consequently, by placing geometric points on each of these three northern continental extremes, the three points can be joined together to form an isosceles triangle with two equal legs measuring 3,142 nautical miles. One leg of this triangle can be extended for half this distance to reach the flank of the highest mountain in North America, Mount McKinley (see figure 9.4).

A direct comparison with this triangle, and its extended line to Mount McKinley, was found earlier from studying Carnac:

an enormous isosceles triangle with legs measuring 6,180 nautical miles that can be drawn around the globe between the summits of Mount Aconcagua, Mount Kilimanjaro, and Mount Elgon . . . one leg of this triangle extends to Mount Everest.*

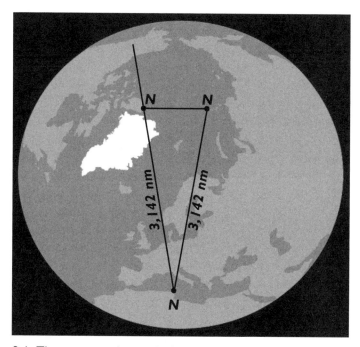

Figure 9.4. The two equal triangle legs are 3,142 nautical miles in length; one leg extends to Mount McKinley. The triangle corners (as described) rest on the northern extremes of Asia, Africa, and North America.

PI AND EVEREST

The precise measurement of 3,142 nautical miles from Cape Angela to Komsomolets finds a point near the southern extreme of the island. From the same point another line measuring 3,142 nautical miles finds the summit of Mount Everest. Consequently, in only three large steps, each measuring 3,142 nautical miles, it is possible to travel from the world's highest point and visit Africa's, Asia's, and America's northern

---

*See the "Continental Summits" section in chapter 8.

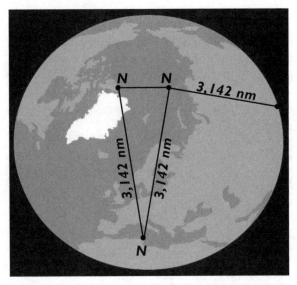

Figure 9.5. Mount Everest and the continental extremes
at Komsomolets, Ward Hunt Island, and Cape Angela;
each line is 3,142 nautical miles long.

extremes. These pi-digit steps are shown in the figure 9.5, with Mount Everest over the horizon.

This unfamiliar aspect of the world's topography emerges as a direct result of joining the points specified by monument alignments. This landscape geometry provides an explanation for the evident Neolithic obsession with marking specific points on the Earth's surface with circles, mounds, and pyramids; Neolithic people were differently wired. Nowhere is this clearer than at the greatest of all the Neolithic high points, the Great Pyramid in Egypt.

GIZA IN THE CROSSHAIRS

Arid, desolate, and mostly uninhabited, Dante Island is about twenty-five kilometers from east to west and twelve kilometers from north to south; it is attached to the African mainland by a sandbar. Dante Island's eastern limit is the most easterly point on the landmass of Africa, slightly farther east than the tip of Africa's horn.

A straight flight from the summit of the Great Pyramid would land on Dante Island's northern peninsula after 1,618 nautical miles. To the south of Giza, more than thirty ancient pyramids were built along the Nile River at Abu Ghurob, Abusir, Saqqara, and Dashur. From the summit of any one of these pyramids, a straight flight to Dante Island would touch down on dry land after 1,618 nautical miles.

Dante Island is specified by this phi-digit measurement from Giza. These two points are therefore joined, and the line is extended. From the southern extreme of Dante Island, a line drawn over the Great Pyramid extends directly to Cape Spear at the eastern extreme of the North American continent (see figure 9.6).

This is the first of two intercontinental lines intersecting at Giza. The second is drawn from Africa's southern extreme at Cape Agulhas to Asia's northern extreme at Komsomolets.

So vast is the canvas that these continental extreme points with Giza in the crosshairs can barely sit in a single view of the Earth.

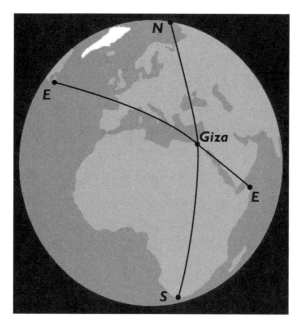

Figure 9.6. Lines intersecting on the Giza plateau can be drawn between two eastern continental corners and a northern and southern continental corner.

As much as we might admire the achievements of the ancient Egyptians, our modern appraisal of their abilities allows no room for conscious or intended alignments of such magnitude. The dual alignment seems to be an unlikely chance event, but whether it is simply a coincidence is debatable. However, there is now an acid test that reveals the unmistakable awareness of those who chose this location. The test, as noted at Avebury, Stanton Drew, Brodgar, Stonehenge, Almendres, Carnac, and so on, is to recognize the aligned points as independent and specified points and to measure between them. The common factors, when these measurements are taken, are pi and phi digits. It is the repetition of these digits, from one site to the next, that illustrates the common code at work. Carnac provided two striking examples of this process, and Giza provides another.

THE ACID TEST

The crosshairs at Giza are created by four continental topographical extremes. These are now joined independently, and the lines are measured:

1. Cape Agulhas to Cape Spear:
   6,282 nautical miles can be measured from one tip to the other $(2 \times \mathbf{3.142} = 6.284)$.*
2. Cape Agulhas to Dante Island:
   3,236 nautical miles can be measured from one to the other $(2 \times \mathbf{1.618} = 3.236)$.

Figure 9.7 illustrates three of the extreme points specified by alignment with Giza joined together.

Couple these measurements with the 1,618 nautical miles from Giza to Dante Island, and the adherence to the common code of practice

---

*To reach the precise tip of Cape Spear, the tolerance required on a line 6,284 nautical miles long is about 1.5 nautical miles. The precise measured line nevertheless goes to the continent's easternmost peninsula.

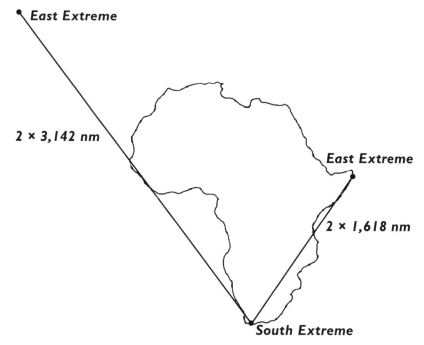

East Extreme

2 × 3,142 nm

East Extreme

2 × 1,618 nm

South Extreme

Figure 9.7. The lines formed by joining the points highlighted by alignment with Giza give pi- and phi-digit multiples when measured in nautical miles.

appears clearly. The greatest of all Neolithic monuments, through simple alignment with topographical extremes, invites us to join the dots. When this is done there is a repetition of pi- and phi-digit measurements, or multiples thereof, between these natural continental extreme points. The same process has produced the same unlikely result on a continental scale at both Carnac and Giza. The Great Pyramid carries this information by specifying topographical points. These points appear to be ordered by measurement.

If 2 × 3,142 nautical miles is measured from Africa's southern corner to Cape Spear, the line passes over the Dakar peninsula on Africa's western extreme and terminates on a chevron-shaped peninsula on the base of the Cape Spear headland at America's eastern extreme. The same base point on Cape Spear is 3,141.5 nautical miles from the summit of Mount McKinley.

In summary, if the distance 3,142 nautical miles is seen as a single rod, it is two rods from Cape Agulhas to the Cape Spear headland, one rod from there to Mount McKinley, half a rod to Ward Hunt Island, a further one rod to Cape Angela, then one rod to Komsomolets, and one rod on to Mount Everest. All the links in the chain are located at continental topographical extremes. This repetition could be seen as something *created* by nature.

10

AMERICAN MONUMENTS

It may be impossible to establish that the ancient world developed a detailed knowledge of the entire planet, but it is relatively easy to demonstrate that numerous monuments were not located randomly in the local landscape. To do this a monument and the surrounding predominant summits are considered as geometric points. If the result of joining the points together is a pattern, repeated from one monument to the next, then the locations were evidently not located randomly.

When this procedure was followed with European monuments, the motif of joined isosceles and right triangles emerged. The accuracy of these geometric figures can be ascertained by measuring the amount of tolerance on the line length that is required to produce a perfect right or isosceles triangle. A computer program can then determine the probability of random points generating triangles to the accuracy found on the ground.*

The computer ran tens of thousands of trials to find the probability. It used two fixed points and introduced a third at random. From this we know with considerable certainty that the probability of the random point creating an isosceles triangle with the two fixed points is around 3 percent when the leg measurements have a tolerance of 1 percent. If the leg

---

*Thanks to Oliver Bentham for writing this program.

measurements have a tolerance of 0.1 percent, the probability of achieving an isosceles triangle reduces to about three chances in one thousand.

The creation of an isosceles triangle between three random points is clearly unusual and cannot be anticipated with any frequency. The creation of a right angle *and* an isosceles triangle from four random points is considerably less probable. For these reasons it is improbable that the location of America's largest pyramid was chosen at random. The location conforms to a geometric motif that has already been seen at Avebury, Stonehenge, Brodgar, and elsewhere, a motif that is demonstrably rare among a cluster of random points.

CHOLULA

Guinness World Records lists the Great Pyramid of Cholula, near Mexico City, as the largest in the world. The pyramid is now covered with vegetation and surmounted by a Catholic church. It is thought that construction began in the third century BCE.

Figure 10.1. The Catholic church on the summit of the Great Pyramid of Cholula. Photo by Jflamenco.

There are three massive natural summits dominating the landscape surrounding the Cholula pyramid; each one is a colossal volcano more than four thousand meters in height, and all are located within forty kilometers of Cholula. These are the second, third, and fifth highest mountains in Mexico, and their summits are plotted along with the Great Pyramid of Cholula in figure 10.2. When geometric points are located on the three natural summits, a right triangle can be drawn between them.

If the mountain summits are considered as fixed points, the computer trials demonstrate that if Cholula were located randomly in relation to these three high points, the probability of creating a right or isosceles triangle with them is about 5 percent. Nevertheless the Cholula pyramid is equidistant between two of these summits, and consequently, an isosceles triangle can be added to the natural right triangle, as shown in figure 10.3.

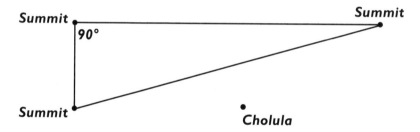

Figure 10.2. Lines joining geometric points located on the three dominant summits overlooking the Great Pyramid of Cholula form a right triangle.

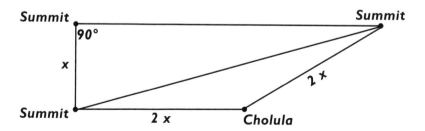

Figure 10.3. The geometry between the natural summits and the Great Pyramid of Cholula summit.

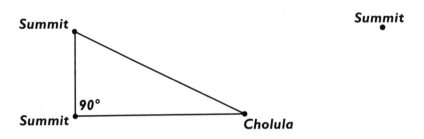

Figure 10.4. Right triangle created using two summit points and Cholula.

Figure 10.3 shows the joined isosceles and right triangles created by joining the high points to the monument with lines. The notion that this geometry was recognized when Cholula was located is further supported by measuring these lines. As the diagram shows the sides of the isosceles triangle are twice the length of one side of the right triangle.

This geometry is created by using a point on the summit of the Cholula pyramid and points on the crater rims of the volcanoes. (These are measurements reminiscent of those made to various points on the perimeter henges in Neolithic Britain.)

A right triangle can also be drawn between the summit of the Cholula pyramid and two further points located on these crater rims, as shown in figure 10.4.

The remarkable ability of the Cholula pyramid builders to locate their own mountain within a geometric pattern created in nature includes a reference to phi. The right angle illustrated in figure 10.4 is tilted 6.18 degrees from the meridian.

The largest pyramid in South America, the Akapana Pyramid at Tiahuanaco in Bolivia, also shows signs of this topographical awareness. The procedure is the same: lines are drawn from the monument to the highest surrounding summits.

TIAHUANACO

The much-dilapidated Akapana Pyramid at Tiahuanaco has a base area of some forty thousand square meters. The site is close to Lake Titicaca

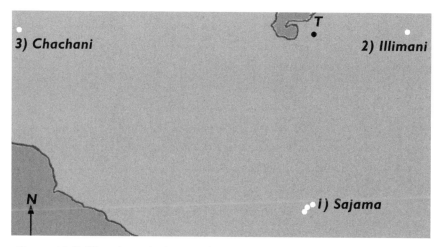

Figure 10.5. The three highest independent points near Tiahuanaco are numbered and named. The highest summit, Sajama is marked 1 and has two outliers. Although higher than summits 2 and 3, these outliers are left out of the following geometry due to their proximity to Sajama. The location of Tiahuanaco is marked by a T.

and is marked by a T in figure 10.5. The image also includes the location of the five highest summits in the area shown.

A right triangle can be drawn between the three named mountains, with the right-angle corner on Sajama. Tiahuanaco is on the hypotenuse of this triangle. A line from Tiahuanaco to Sajama divides the first right triangle into two further right triangles.

The Akapana Pyramid at Tiahuanaco is located a few hundred meters from the triangle hypotenuse, line 2 to 3. The point marked 2 is on the southern summit of Mount Illimani. At both Cholula and Tiahuanaco, the regional high points surrounding the monument create a right triangle when they are joined together. Moreover, the motif, seen at Cholula, of a right triangle and an isosceles triangle joined on one side is also repeated at Tiahuanaco. When the fourth and fifth highest regional summits are included as geometric points, an isosceles and an equilateral triangle can be drawn, with both triangles sharing one side. All this geometry is dependent on the location of Tiahuanaco as a geometric point.

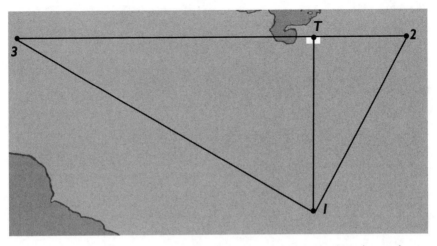

Figure 10.6. When the three natural high points are joined, a right triangle can be drawn between them.* Tiahuanaco is on the hypotenuse, and by joining Tiahuanaco with the highest summit (1), the first right triangle is divided into two further right triangles.

The same geometric motif found at two of North and South America's largest ancient sites is produced again at perhaps the most famous of them all, Machu Picchu in Peru.

MACHU PICCHU

The Cusco Region of the Andes has hundreds of craggy peaks to choose from, so why was this particular precarious perch singled out for the monumental architecture of Machu Picchu?

In local legend the highest summit in this region, Salcantay, was wedded to the second highest summit, Veronica, a beautiful mountain some thirty kilometers away from its partner. These are the two highest summits in the region of Machu Picchu. Mount Salcantay is due south of Machu Picchu. Mount Veronica is due east of Machu Picchu.

The remote and ancient complex is therefore located on the right-

---

*The south summit of Mount Illimani is adopted as the geometric point; the right angle is then found above the snow line on Sajama's summit.

Figure 10.7. Machu Picchu, precariously located among
mountain summits. Photo by Allard Schmidt.

angle corner of a triangle created by lines extended in cardinal directions to the two regional high points.

Figure 10.8 shows this right triangle aligned to the meridian, but it also includes the highest summit to the west of Machu Picchu, Mount Sacsarayoc.*

When this third high point is included in the geometry, the motif of a joined right and isosceles triangle is found once again; the two joined triangles, aligned to the meridian, are shown to scale in figure 10.8.

---

*There is some doubt about the name of this peak. It is found by using the relief map facility on Google Earth.

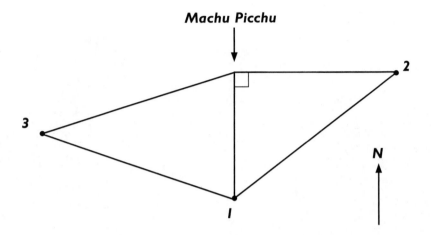

Figure 10.8. The isosceles and right triangles formed by joining points 1 (Mount Salcantay), 2 (Mount Veronica), and 3 (Mount Sacsarayoc) with Machu Picchu.

Using Google Earth the right-angle corner and the isosceles corner share a point about two hundred meters from the entrance to the site. This point is ten nautical miles due north of the summit symbol on Mount Salcantay. Heading due east from this point for 12.72 nautical miles, the line terminates on Mount Veronica, at a point about two hundred meters from the summit. From this point it is 16.18 nautical miles back to the summit of Mount Salcantay. So, allowing only small tolerances, it is possible to draw a right triangle with the proportions of phi expressed in nautical miles between the length of the line following the meridian and the length of the hypotenuse.

Even with porters to carry their bags, today's adventurers to Machu Picchu all agree: the trail is steep, the air thin, and the weather fickle. Yet the builders of Machu Picchu laboriously dragged boulders to a location at a right angle to the two highest peaks *by chance*. At Tiahuanaco and Cholula an almost identical *coincidence* occurred. Without rhyme or reason the monument builders in each case located their monuments at right angles to the surrounding highest summits. At Machu Picchu the right angle is also aligned to the meridian. In each case isosceles triangles are attached to the right triangles. The use of *chance* or *coincidence* to explain the repetition of this geometry is not realistic. Nevertheless, the current orthodox view prevails: our ancestors were too ignorant to achieve what they evidently did achieve. From among the hundreds of equally precipitous peaks, the builders chose the location of Machu Picchu with little regard for, even less understanding of, and certainly no measurement of the surrounding topography. We need another explanation.

The monuments are expressing something fantastic. Each monument is describing a geometrically ordered topography.

When the computerized probabilities are considered, the analyst who insists that the monuments at Cholula, Tiahuanaco, and Machu Picchu were randomly located is, to use an idiom from cricket, batting on a dodgy wicket. The probability that the locations were chosen randomly is extremely small. In fact the probability is almost too small for sensible consideration when the same unusual geometric motif is recognized at one site after another.

The same pattern is found once again, this time at the most renowned pyramid complex on the North American continent, Teotihuacán.

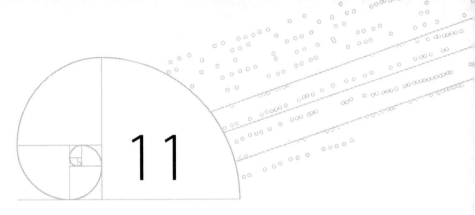

COPYING NATURE

TEOTIHUACÁN

City of Negatives

Like many other ancient sites, Teotihuacán is remarkable for what is not known about it. Nobody knows with any degree of certainty who built it. The Totonacs or the Mayans may have played a part, though neither apparently had metal tools, pack animals, or the wheel. Described as "the largest and most remarkable metropolis ever to have existed in the pre-Columbian New World,"[1] Teotihuacán's rise was as monumental as the buildings that characterize it. Said to have had a multiethnic population and to have been home to the Nahua, Zapotec, Otomi, and Mixtec people, it appears to have grown from virtually nothing to a sprawling metropolis in the first centuries AD. The reason for building pyramids at this location remains mysterious.

Teotihuacán's decline is variously attributed to a cataclysmic fire or to repeated sackings and burnings, but by whom, and why, also remains mysterious. Although murals are plentiful the people who built Teotihuacán left no written records, and the site was named by the Aztecs many years after its demise. The name Teotihuacán roughly translates as "the place where men become gods."

Figure 11.1. The Pyramid of the Sun photographed from the Pyramid of the Moon. The two pyramid summits fall into alignment with the natural high point on the horizon. Numerous small step pyramids are seen in the foreground. Photo by SElefant.

Two massive pyramids overlook Teotihuacán: the Pyramid of the Sun, said to be the third largest in the world, and its neighbor, the Pyramid of the Moon. The photograph in figure 11.1 was taken from the Pyramid of the Moon looking at the Pyramid of the Sun. The two man-made high points align with the highest point on the horizon about seven kilometers from the pyramid of the Moon: ^-^-^.

The summit in the distance in figure 11.1 can be seen as a specified point because it aligns with the two man-made pyramids. But there is a second natural high point called Cerro Gordo that is a little closer to the Pyramid of the Moon.

The Pyramid of the Moon was built up gradually to its present size. Excavations by Saburo Sugiyama and Ruben Cabrera have shown that

Figure 11.2. The Pyramid of the Moon viewed from the Avenue of the Dead, with Cerro Gordo looming about five kilometers away. Photo by Antony Stanley.

the pyramid achieved its current dimensions (168 × 149 × 46 meters) over several centuries, involving seven separate building stages. The Sun and Moon pyramids at Teotihuacán are truly monumental, and additional smaller step-pyramid structures are plentiful. Tourists today can walk along the so-called Avenue of the Dead, starting from the entrance to the Cuidadela (a sunken plaza surrounded by fifteen step pyramids), and follow the ancient causeway, flanked by further step pyramids for two kilometers, along a straight line to arrive at the base of the Pyramid of the Moon. The causeway is aligned to the pyramid summit, and the line can be continued over the Pyramid of the Moon to the revered natural mountain Cerro Gordo, which is the highest summit in a twenty-kilometer radius of the site. The avenue aligns within a few hundred meters of Cerro Gordo's summit.

In addition to Cerro Gordo there is a single nearer isolated summit to the west of the pyramid complex.

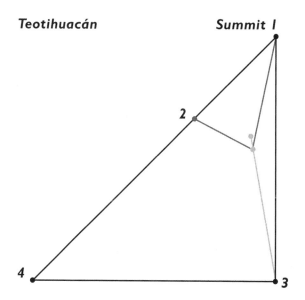

Figure 11.3. Geometry between high points surrounding Teotihuacán. An isosceles triangle is outlined; a right-angle isosceles triangle is also outlined. The two pyramid summits are indicated by dots on the gray line.

On the plan in figure 11.3, the two local summits are marked 1 and 2. When these two summits are plotted, along with the summit platform of the Pyramid of the Sun, an isosceles triangle can be drawn between them. This triangle is outlined in figure 11.3.

In figure 11.3 one leg of the isosceles triangle extends to a third natural high point, Cerro Chiconahulta (point 4), arriving at the base of this volcanic hill ten nautical miles from Cerro Gordo. The three high points are aligned from summit to summit to base. Summit 2 is found at the phi interval on this line. This ten-nautical-mile line is the hypotenuse for a right-angle isosceles triangle. The right-angle corner of this triangle rests on summit 3, which is the summit aligned with the Pyramid of the Sun and the Pyramid of the Moon (on the gray line).

In summary, allowing tip-to-base measurements, the result of joining the natural specified points at Teotihuacán is a right-angle isosceles triangle. The hypotenuse of this triangle is 10.00 nautical miles in length, and summit 2 is located at the phi interval on it.

Figure 11.3 illustrates many of the familiar keynotes: monumental high points are aligned with a natural high point, a right and an isosceles triangle are created, and a reference to phi and the nautical mile is coupled with tip-to-base measurements. But these local measurements at Teotihuacán only hint at the greater reason for choosing this extraordinary location.

At other ancient sites the digits 3142 and 1618 appear with unaccountable regularity when measuring between a monument and landmass extreme points. Could the same be expected at Teotihuacán?

THE MIRROR OF MOUNT EVEREST

1. Measured on Google Earth the distance between the Pyramid of the Sun at Teotihuacán and the northern contiguous landmass extreme of North America, Zenith Point, is 3,142.7 nautical miles.*
2. In Asia 3,142 nautical miles can be measured from Mount Everest to the northern extreme island of Komsomolets.
3. Both lines can be drawn with a bearing of 1.618 degrees.†
4. Teotihuacán duplicates, in North America, the location of the world's highest mountain a continent away in Asia.

If Komsomolets and Zenith Point are both recognized as continental corners, then it is true to say that lines bearing 1.618 degrees and measuring 3,142 nautical miles from Teotihuacán and Mount Everest both terminate on northern continental corners.

---

*Contiguous refers to the unbroken landmass of the continent. Zenith Point is the northern limit of the landmass, as opposed to Ward Hunt Island/Cape Columbia at the northern limit of the continent.
†A line 3,142 nautical miles long bearing 1.618 degrees taken from the Pyramid of the Sun terminates at a base point, a finger of land on Zenith Point, one nautical mile from the precise northern tip.

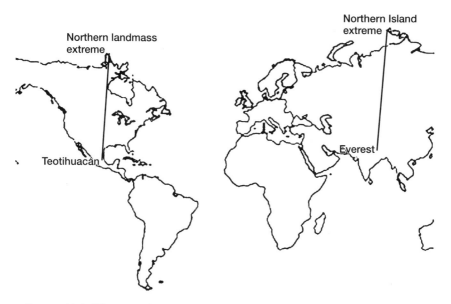

Figure 11.4. The two lines in the image are distorted by the Mercator projection but both are 3,142 nautical miles long, and both have a northerly bearing of 1.618 degrees. One extends from Mount Everest to Komsomolets, the other from Teotihuacán to Zenith Point.

ZENITH POINT

In current historical records Zenith Point on the Boothia Peninsula remained uncharted until the first half of the nineteenth century, when Captain John Ross and his crew spent three grueling ice-bound years there on the paddle steamer *Victory*. They failed in their attempt to reach the western side of Boothia by sea and thereby achieve their goal of being the first to sail the elusive Northwest Passage. Having abandoned ship they suffered unimaginable hardship in the clutches of some of the worst cold-weather conditions this planet has to offer, and their safe return owed as much to good fortune as to their resolve.

Ross returned to London still unaware that Boothia was the northernmost peninsula of mainland North America and not, as he had imagined, an island. The peninsula reaches its northern landmass extreme at Zenith Point, overlooking the treacherous Bellot Strait, named after Joseph Rene Bellot, who traveled there with Captain William Kennedy

Figure 11.5. J. Branard's copy of a picture made by Captain John Ross aboard the *Victory* in 1835 showing an Inuit man (an Inuk) drawing a map, or following a chart, of the Boothia Peninsula.

on another grim voyage. Bellot and Kennedy were the first Europeans to see and recognize North America's northern landmass extreme in 1852.

This is Western history's earliest record of the discovery of the northern point on the continental landmass. But the local Inuit were aware that Boothia was a peninsula (and therefore had no south coast); they told Captain Ross as much and even made appropriate alterations to his erroneous charts. Ross, however, did not trust the natives' judgment. How could the Inuit know where they were on a map?

The distance and bearing from Teotihuacán to Zenith Point can be seen as a form of numerical language. The location of Teotihuacán communicates through the digits 3142 and 1618 when measuring a line to the northern landmass extreme (exactly the same process adopted at Almendres). On the continent of South America, the northern peninsula appears to be identified in a similar manner.

GUAJIRA

The Guajira Peninsula is the large body of land (over one hundred kilometers from east to west and almost surrounded by the sea) on the northern extreme of South America. Measuring 1,618 nautical miles from the Pyramid of the Sun finds the base of a finger of land (about five kilometers in length) located on the northern coast of the Guajira Peninsula. Following the direction of this finger leads to the precise northern extreme of the Guajira Peninsula.

In this way pi- and phi-digit measurements can be made from Teotihuacán to the peninsulas at the northern landmass limits of both North and South America. There are over 35 million square kilometers of land on the North and South American continents and only a few square kilometers from where both these measurements can be taken. The pyramids of Teotihuacán occupy this small area.

To test the probability of finding this location by chance, measurements were taken from a random virtual tour of cities: Houston, Austin, Albuquerque, Santa Fe, Phoenix, San Francisco, Los Angeles, Monterey, San Antonio, Birmingham, Miami, Augusta, Charlotte, New York, and Portland. Not one of these cities is 3,142 nautical miles from a continental extreme point, nor are any 1,618 nautical miles from a continental extreme point. Random locations in America do not frequently appear at either of these two distances when measuring to topographical extremes.*

In early studies of this beautiful complex of monuments at Teotihuacán, we had not fully understood the code of practice, nor could we quite believe that European Neolithic ideals could possibly be translated three thousand years later in America. So we arrived in America with two repetitive Neolithic measurements defining topographical extremes, and against seemingly ridiculous odds, Teotihuacán deftly illustrated the use of both of them. Coincidence is a highly improbable explanation. A more reasonable explanation is that despite

---

*The cities were chosen randomly, and the measurements were taken from the "location dot" produced by Google Earth. Measurements were taken to all the cardinal landmass extremes of the American continent. The closest result was ten nautical miles off target.

their spatial and temporal isolation, many of the major monuments around the world were located by people with common ability, motivation, and understanding. At Teotihuacán, as with Stonehenge, there is no way of knowing who first marked the location, nor when that marker was created, but an understanding of why these locations around the world were so important is beginning to emerge. We may never understand how the builders achieved this, but the measurements speak for themselves. With only minimal tolerances on these lines, Almendres is 314.2 nautical miles from the northern Iberian landmass extreme on a bearing of 3.142 degrees. Teotihuacán is 3,142 nautical miles from the North American landmass extreme on a bearing of 1.618 degrees, and so on.

Teotihuacán follows another alignment principle found in Neolithic Europe. In all cases the monuments are located on lines joining at least two topographical extreme points. Teotihuacán is no exception; the scale is global, just as it is at Carnac.

Teotihuacán is on a line joining the western limit of South America, at Punta Balcones*, and the western limit of the Alaskan Peninsula landmass, Teotihuacán, is located one-third of the way along this line measured from the north side of Punta Balcones.

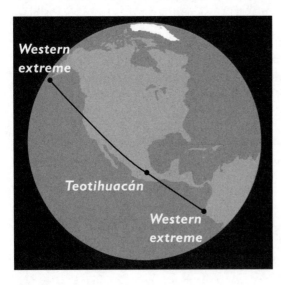

Figure 11.6. With a line drawn from the western landmass extreme of South America to the western landmass tip of the Alaskan Peninsula, Teotihuacán is one-third of the way along the line.

*The notice board on the beach says Punta Balcones, but others call this headland Punta Parinas.

The western tip of the Alaskan peninsula is not the most westerly extreme of the North American continent. The extreme western point is at Cape Prince of Wales on the Seward Peninsula over one thousand kilometers to the north. From a base point on this horn-shaped peninsula, a line can be drawn over the summit of Mount McKinley to Teotihuacan. The monument, the mountain, and the cardinal extreme point are thus aligned.

THE SEWARD PENINSULA IN ALASKA

Cape Prince of Wales, on the Seward Peninsula, is at the extreme western limit of the contiguous landmass of America, and the city of Nome sits on the peninsula's south coast. The inhabitants are well accustomed to extreme conditions; the port is frozen from November to early summer, which is beautiful but brief. Above the hillside cemetery, its white crosses bent by the wind, seabirds wheel and squeal above the Bering Strait; musk ox, moose, and grizzly bears roam the hinterland, and migrating swans and cranes herald the onset of winter. A promotional video, with a backdrop of native fauna, tells us, "There's no place like Nome. . . . The locals are wild, and we think you'll fit right in."

On January 22, 1925, a desperate Curtis Welch, the only doctor in Nome, sent a radio telegram:

> An epidemic of diphtheria is almost inevitable here STOP. . . . I am in urgent need of one million units of diphtheria antitoxin STOP.

Enter man's best friend. With planes dismantled for the winter and other routes impassable, dog teams braving an overland mail and supply route were the community's only choice. Led by the husky Balto, a team arrived in Front Street eleven days later with the first precious batch of serum. Today the husky's statue stands in Central Park, a lasting tribute to the relay teams and their mushers, who endured subzero temperatures, headwinds of hurricane force, whiteouts, icy waters, and, in one case, an entanglement with a herd of reindeer to complete what is arguably one of the greatest and record-breaking feats of man and beast. Figures vary slightly, but the general consensus is that in total the relay teams and their drivers had taken around 128 hours to cover some 674 miles. Another great lead dog, Togo, "the sled dog overlooked by history," led his team and his musher, Leonhard Seppala, on a mind-boggling 260-mile loop, on an outward- and then Nomeward-bound journey. Togo now stands—in stuffed and mounted glory—at the Iditarod Race Headquarters in Alaska. Every year the Burled Arch is erected on Front Street in Nome to greet the first musher and team to finish this incredible commemorative race. If you want to be there to cheer them home, book your hotel room early.

A UNIVERSAL MOTIF

Remote and lonely though the continent's western extreme may be, the tip of the Seward Peninsula has a part to play in the landscape geometry of Teotihuacán.

The people who recognized the locations for these monuments speak with one voice: "Here is a point where the geometric motifs of the planet can be found." It seems the status of a location was based on the quantity and quality of alignments, including phi- and pi-digit measurements that could be made from a small area on the ground. In turn the points specified by this process should then reveal geometry or order on the globe. In questioning how these calculations could be achieved by people in prehistory, we come to understand the questions that baffled Professor Alexander Thom at Avebury. There he was amazed by the accuracy and complexity of Neolithic measurements; at Teotihuacán the same refined ability is clearly evident and is equally inexplicable. The problem lies less with the evidence than with the willingness to accept it as evidence; finding it beyond a reasoned understanding of history, we tend to fall back on "coincidence" or "chance" to explain it. But these comfortable options make no sense. Having already found an ancient code that adopted pi- and phi-digit measurements in Europe, surely the accuracy of the 3,142-nautical-mile measurement from Teotihuacán to Zenith Point rules out coincidence in a reasonable argument? Yet in orthodox circles the accepted view of history still demands that all this is a coincidence. Adherence to an unsustainable doctrine continues today to prevent students from enjoying the full riches of our marvelous history. It was this same intransigence that obstructed Thom and many others who have tried to open our eyes to ancient wisdom. The monuments cohere to a common global pattern; they are vehicles transporting esoteric information. Should it just be ignored?

Be it Carnac, Giza, or Teotihuacán, these monuments all present the globe in the context of a geometrically organized object. The same message is repeated over and over again. It describes a real sense of order on the planet that is revealed by joining the corners of the landmasses together. When this process is applied to the two largest landmasses on the Earth, they appear to confirm the truth of the message.

A SENSE OF ORDER

The sense of geometric order on the Earth emerges on a grand scale in Asia. The northern and southern limits of the Asian landmass share the same degree of longitude. Other than coincidence there is no valid explanation for any large landmass to have its southern and northern corners so closely aligned with the two poles of the Earth, yet the Earth's largest continent, spanning nearly 200 degrees from east to west, does so. But the coincidence is increasingly mysterious because the same meridian alignment applies to the continent of South America. Both meridians are shown in figure 11.7.

The Asian meridian is taken from the southeastern fingertip of the Malaysian Peninsula; this passes to the tip of Cape Chelyuskin and over the North Pole. It continues to the North American continent and passes within twenty kilometers of Ward Hunt Island.

Figure 11.8 shows the Earth viewed from above the North Pole. The vertical line passing over Africa's northern and southern extremes and continuing on to Mount McKinley is intercepted by lines from the

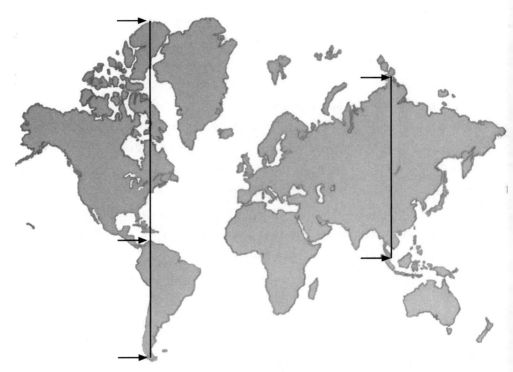

Figure 11.7. The lines joining the northern and southern landmass extremes of South America and Asia appear as two parallel lines on a Mercator map. The South American meridian, from Cape Froward to the Guajira Peninsula, continues to the North Pole, passing over the northern extreme of North America at Ellesmere Island near Cape Columbia.* The continental extremes are marked with arrows.

left and right, joining the northern and southern extremes of Asia and South America. These near-horizontal lines do not intercept the vertical line at exactly 90 degrees, but they do illustrate an interesting focal point in world topography—Ward Hunt Island.

The lines joining these northern and southern corners of Asia, America, and Africa convey a sense of order that becomes more pro-

---

*The meridian of Cape Froward in the south passes over the Guajira Peninsula and continues to the flank or base of the landmass on which Cape Columbia is located. The two precise northern and southern limits are just over one degree of longitude apart.

nounced when a further line is drawn between continental extremes, this time joining the eastern and western extremes of the Asian landmass. The additional line is shown in figure 11.9.

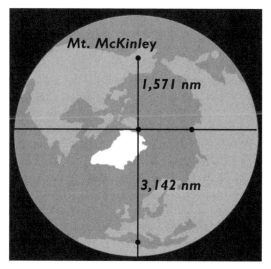

Figure 11.8. The lines shown on this map pass over northern and southern continental corners of Asia, Africa, and South America. The lines cross near Ward Hunt Island. The line from Africa is extended past Ward Hunt Island to Mount McKinley.*

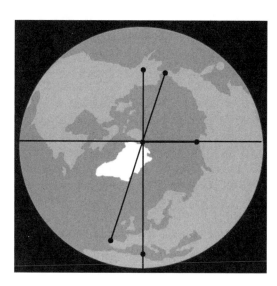

Figure 11.9. A repeat of figure 11.8, with an additional line joining Eurasia's eastern and western landmass extremes, at Cape Dezhnev and Cabo de Roca.

---

*On the scale in figure 11.8, the line joining Asia's northern and southern peninsulas appears to continue to the northern and southern peninsulas of America. A closer view reveals two lines meeting near Ward Hunt Island at an angle of about 176 degrees.

THE QUINCUNCIAL MAP

The cardinal corners of landmasses used to create the image in figure 11.9 are all considered to be randomly distributed. But another perspective could recognize these extreme continental points as ordered. The quincuncial map, shown in figure 11.10, illustrates the same alignments in flat view. The focal point is Ward Hunt Island: all lines pass within ten nautical miles of it.

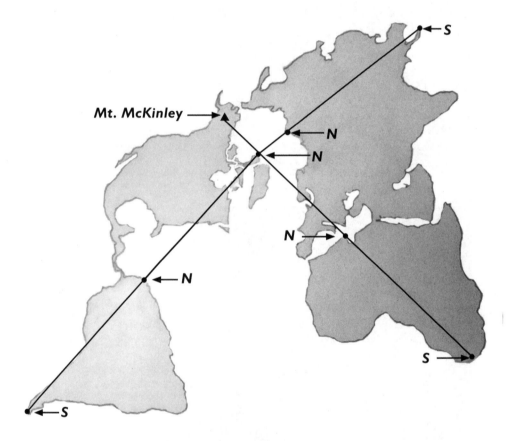

Figure 11.10. Quincuncial map showing lines joining northern and southern extremes of Asia, Africa, and America (indicated with arrows). The map includes Mount McKinley, but (for clarity) excludes the line joining Asia's eastern and western extremes, which also passes through the focal point near Ward Hunt Island. The focal point has a radius of about ten nautical miles.

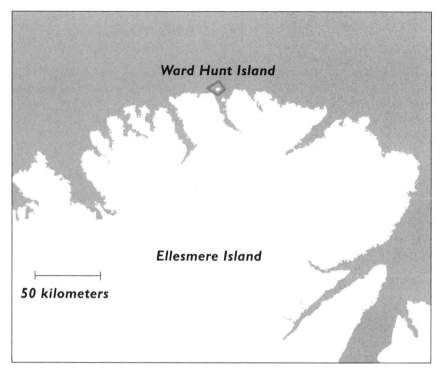

Figure 11.11. The northern reaches of Ellesmere Island at the northern limit of America; Ward Hunt Island is within the kite shape. All the continental lines joining the continental corners meet within the kite.

Figure 11.11 gives a sense of perspective. It shows the northern limits of Ellesmere Island at the North American continent's northern extreme. Ward Hunt Island is within the kite shape; all the continental lines meet within this outlined area.

The small area surrounding Ward Hunt Island is aligned with so many continental extreme points that we became curious. Would a line from Mount Everest passing over Ward Hunt Island result in a three-point alignment with a monument?

It does. The line extends to St. Louis, Missouri, once called Mound City and home to the largest ancient earth mound on the North American continent.

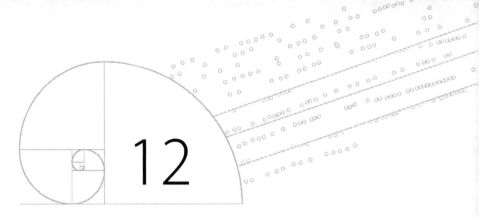

12

COLD CONTINENT

CAHOKIA

St. Louis, Missouri, on the Mississippi River, was once known as Mound City for good reason. Settlers, arriving largely from Europe, were confronted with ancient earthworks of unparalleled proportions in this fertile, flat lowland region. They found the horizon broken by hundreds of earth mounds covering an area of approximately four thousand acres. Several of these were enormous structures; one, Monk's Mound near East St. Louis, is the largest in North America, with a base area of four acres (sixteen thousand square meters). These mounds were built around 1000 CE, when the Khmer kings of Angkor were building huge stone pyramids 7,500 nautical miles away in Cambodia.

In both cases armies of workers were required for construction, and in both cases many of the buildings were solid throughout and had little if any utilitarian value, except perhaps an improved view. At Cahokia and Angkor alike, vast quantities of material were piled up into structures where the function appears to have been little more than to create high points in the landscape. Twenty-two *million* cubic feet of soil were shifted to make Monk's Mound in Cahokia, about twice the volume of Silbury Hill in England. The tiered mound has been likened to a step pyramid, and it overlooks a floodplain where a mound of similar size, ten kilometers distant, was destroyed fewer than two hundred years ago. The great mounds of earth at Cahokia gave teams of railway workers a ready supply of bedding material for railroad construction. Monk's Mound is one of the survivors.

Archaeologists have varying views on the people responsible for Cahokia; there is no consensus on where this culture came from or why they built mounds. After about three hundred years, these people dispersed and left no written word. This lack of any written record is a recurring feature among the monument builders at Teotihuacán, Giza, and Neolithic Europe. But the enduring common legacy from these cultures is dramatically altered topography. Their monuments can be seen as huge statements in their own right, writ large on the surface of the Earth and apparently revealing a common understanding. And the key to their monumental language is measurement.

MEASURING CAHOKIA

A straight line joins Monk's Mound with two South American extreme points. The line is drawn from Monk's Mound to Horn Island, the location of Cape Horn, traditionally seen as South America's southernmost point. The line from Monk's Mound to Horn Island passes over Punta Balcones, the peninsula at the western extreme of South America. Again the monument aligns, ^-^-^, with two continental extreme points. Could this be a coincidence? When the line is measured, the code is tested. Sure enough 3,142 nautical miles separate the two specified continental extremes. The precise measurement from the tip of Punta Balcones extends for 3,142 nautical miles to the center of Horn Island; the line drawn from Cahokia is shown in figure 12.1.

The mounds at Cahokia draw attention to this "pi distance" in nature by the same methods of alignment and measurement found at Teotihuacán and among the Neolithic monuments. This pi-digit distance has already been found separating the northern extremes of Asia and America from Africa, and it was recognized by following the alignments at Carnac. The upshot is the realization these two cardinal

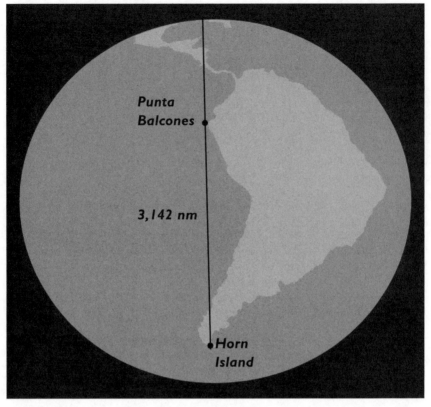

Figure 12.1. The distance between the dots is 3,142 nautical miles. The line continues northward to Monk's Mound at Cahokia.

corners of South America can be seen as part of a mysterious common measurement defining continental limits. This understanding is expressed at Cahokia in a language adopted thousands of years earlier in Neolithic Europe. And like their Neolithic counterparts, the monuments at Cahokia repeatedly express an allegiance to pi- and phi-digit measurements.

For example, a line bearing 314.15 degrees from Monk's Mound extends to the very tip of the contiguous land on the Alaskan Peninsula. In turn the line joining this Alaskan peninsula point with Punta Balcones passes over Teotihuacán, and 3,142 nautical miles can be measured between Punta Balcones and Horn Island on a line extending to Monk's Mound. It is a joined-up language connecting the limits of the

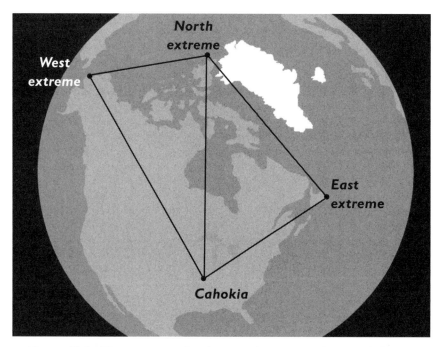

Figure 12.2. The two joined right triangles appear distorted due to the curvature of the Earth. The right-angle corners are on Cape Spear (East extreme) and Ward Hunt Island (North extreme).

continents. These monuments are geometric points defining the vision of a geometrically ordered landscape.

A line joining Ward Hunt Island to Monk's Mound proves to be the hypotenuse of a right triangle with America's eastern extreme, Cape Spear, at the right-angle corner. This is a 90.00-degree angle measured using Google Earth's ruler and island outlines. The three corners of the triangle are located on Monk's Mound, the eastern limit of America at the tip of Cape Spear, and the eastern limit of Ward Hunt Island. A second right angle can be measured between Monk's Mound and Ward Hunt Island, but this time the western peninsula of America is on the third corner; the precise geometric point is found on the flank of the Seward Peninsula. The position of Cahokia thus becomes unique and slightly magical. It illustrates how to create geometry with continental corners (figure 12.2).

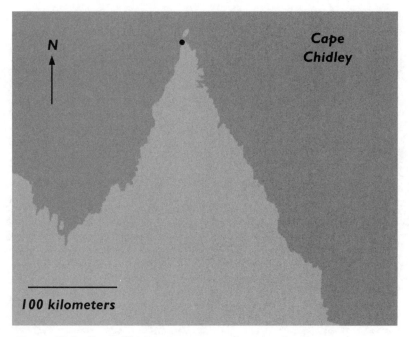

Figure 12.3. Cape Chidley in eastern Canada; the dot at the apex
is 1,618 nautical miles from the Monk's Mound at Cahokia.

AN EPIC JOURNEY

One further measurement from Cahokia led to an epic journey around
the coast of Greenland. A common distance in the code is 1,618 nautical miles; measuring this distance from Monk's Mound takes us to
the tip of another topographical extreme, the distinctly chevron-shaped
Cape Chidley, as shown in figure 12.3.

The dot on Cape Chidley is 1,618 nautical miles from Cahokia.
From this point lines drawn to the southern and western extremes of
Greenland form a right angle, as shown in figure 12.4.

The right angle shown in figure 12.4 inspired a closer examination of the corners of Greenland, which is the world's largest island,
but being less than one-third the size of Australia, it is not big enough,
by modern criteria, to be defined as a continent. Greenland is also part
of the North American tectonic plate, a subject of some interest to the
adventurous Alfred Wegener.

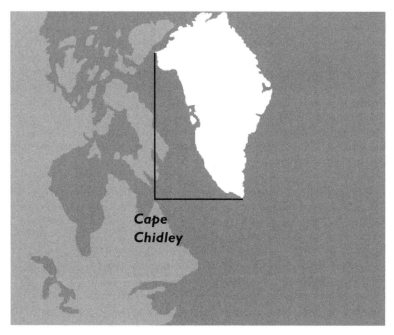

Figure 12.4. A right angle can be drawn between the southern and western extremes of Greenland, with the tip of Cape Chidley at the right-angle corner.

WEGENER IN GREENLAND

The meteorologist Alfred Wegener (1880–1930) was one of the first Westerners to spend a winter on the ice sheets of central Greenland. At the age of fifty, he lost his life on a grueling sled run, having resupplied the camp at Eismitte ("Ice Middle"), rather unsurprisingly located in the heart of the island. His colleague on the trip was a twenty-three-year-old Greenlander, Rasmus Villumsen. The body of Wegener was buried carefully in sown-up sleeping bags lying on a reindeer skin in the frozen wilderness 150 kilometers from the coast. Upright skis had been used to mark the burial, and they were found some six months later. But his companion, Villumsen, was never seen again.

Eismitte is more than three hundred kilometers from the nearest coast; any temperature rise above zero centigrade would be regarded as a fine, mild day in a region where −64 degrees centigrade has been recorded. It is an unforgiving location, as Wegener discovered, but in his case the world's largest island was perhaps an appropriate place to die.

By the 1920s the multitalented Wegener had already proposed in his book, *The Origin of Continents and Oceans,* that the continents of the world are prone to drift and to shift about over long periods of time. In the scientific climate of the day, it was an outrageous suggestion, especially coming from a mere weatherman. Wegener found no shelter from the disparaging responses to his idea. "Utter, damned rot!" the president of the American Philosophical Society blustered. One British geologist suggested that anyone who "valued his reputation for scientific sanity" would never dare support Wegener's preposterous theory. Wegener's idea was described as a fairy tale of "geopoetry" by one animated critic, and the American Association of Petroleum Geologists organized an entire symposium with the primary purpose of debunking Wegener's ideas.

By the mid-1960s, Wegener's theory was generally accepted as correct; plate tectonics became well accepted. It was a long thaw. More than thirty years after his death, Wegner was finally recognized as the father of plate tectonic theory, but this currently accepted theory is now being challenged as further information about shifting seabeds is gathered.

A SHIFT

Wegener has been compared with Galileo because he directed his contemporaries toward a paradigm shift. He provides just one of innumerable examples of how difficult it is to alter accepted points of view. It would appear that there is something in human nature that causes us to balk at the idea of changing our minds. Galileo spent the last eight years of his life under house arrest by order of the Catholic Church, his discoveries having failed to sway the minds of his critics. Although protesting that he had kept an earlier command about holding or disseminating any "condemned opinions," he was forced to admit that readers of his *Dialogue Concerning the Two Chief World Systems* could perceive his Copernican leanings.

The discovery of x-rays in 1895 provides another example of how a new phenomenon "violated deeply entrenched expectations."[1] The x-ray appeared to be paranormal at the time it was discovered and was initially described as "an elaborate hoax" by Lord Kelvin.

The Gaia theory proposed by James Lovelock is a more recent example of human intransigence. Lovelock suggested that the Earth possessed a self-regulating system. At the time the idea was simply too religious; it appeared to require something omnipotent; it was beyond the realms of contemporary understanding. No scientific journal would publish the theory for twenty years. It is now mainstream science.

RIDICULOUS CRYSTALS

More recently Dan Shechtman was ostracized from the scientific community and called "a disgrace" by his peers for the discovery of quasi-crystals. These crystals were described as "ridiculous" at the time because they displayed previously unknown fivefold symmetry. They do, however, exist, and their structure is based in phi proportions.

Shechtman won the Nobel Prize for his discovery in 2011.

Shifting planets, shifting landmasses, regulated planets, unbelievable x-rays, and "ridiculous" crystals: these are only a few of many cases, all of which highlight a persistent failure to recognize a change in perspective or even to recognize the persistent failure.

If the people of the ancient world understood the planet to be ordered in some mysterious way, on what premise have we assumed that they just imagined the whole thing? And is that premise valid?

MEASURING GREENLAND

The highest summit in Greenland is the giant Gunnbjørn Fjeld, a beautiful pyramidal peak, the highest summit inside the Arctic Circle. The mountain is located on Cape Brewster, the large horn-shaped peninsula on Greenland's eastern side.* At the very tip of Cape Brewster is an isolated triangular lump of rock about ten nautical miles across. A line bearing 6.18 degrees from the mountain summit passes to one corner of this rocky limit.

From geometric points located on this small triangular rock island, it is possible to measure the same distance to the northern and to the southern peninsulas of Greenland. So an isosceles triangle can be drawn with corners located on these three landmass limits.

The tip of Cape Chidley also has an isolated rock island off its northern extreme. This rock marks one corner of an isosceles triangle with the northern and southern corners of Greenland. These are the same northern and southern points that are equidistant from the tip of Cape Brewster. The overall result is two isosceles triangles and a right angle drawn between the cardinal extreme peninsulas on Greenland and points located on the landmass and island extremes of Cape Chidley; see figure 12.5.

The idea that the world was once seen and understood in this measured and geometric way in antiquity is understandably contentious. But the geometry shown so far in this book resembles the "fashioning," the "ordering," and the "designing" spoken of repeatedly in the ancient world. If geometry persists between topographical extreme points, then is it not perfectly legitimate to assume the points themselves are not randomly located? A pattern requires formulaic repetition, and the repetition of phi- and pi-digit measurements between topographical extremes satisfies this requirement. It is 618 nautical miles from the northern extreme of the islands off Cape Chidley to the southern extreme peninsula of Greenland, on Egger island; the line terminates at the base of the southernmost peninsula.

According to the description given by Hermes Trismegistus in the *Hermetica,* this fashioning could be seen "everywhere" on Earth. And

---

*Cape Brewster appears as the prominent eastern point on Greenland; however, due to the closing of lines of longitude, the true eastern extreme of Greenland is in the north, near the northern extreme.

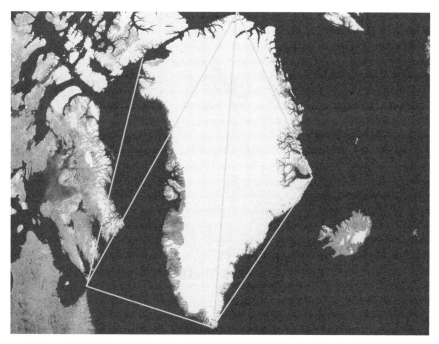

Figure 12.5. Satellite view showing the geometry between cardinal extreme points on Greenland and Cape Chidley: two isosceles triangles and a right angle.* Photo by NASA World Wind.

the ancient monuments repeatedly specify landmass extremes by aligning with them. By simply joining geometric points located on the landmass extremes of Greenland, a sense of measured geometric order on the Earth becomes visible in figure 12.5. This geometry is coupled with the repetition of pi- and phi-digit measurements that further reinforce the understanding that these corners of Greenland are not randomly scattered and a form of order is dictating their locations. The isolated lump of rock on the tip of Cape Brewster in Greenland is also 1,618 nautical miles from Cape Spear, the eastern extreme of North America.

Does this eastern corner of North America appear to be geometrically ordered with the other corners of the North American continent?

---

*The right-angle corner is on the tip of Cape Chidley; the isosceles triangle corner is on the small island just off the peninsula.

AN AMERICAN KITE

Cape Spear and three further "corners" of the North American continent provide another example of how landmasses can appear to be geometrically ordered when the corners are joined together.

Figure 12.6 shows a perfect cross (four right angles) forming a kite with the four outer corners resting on four of North America's most extreme points.

In figure 12.6 the eastern geometric point is on Cape Spear. In the south the point is at the southern end of the Baja California Peninsula in Mexico. In the north the point is on Ward Hunt Island. In the

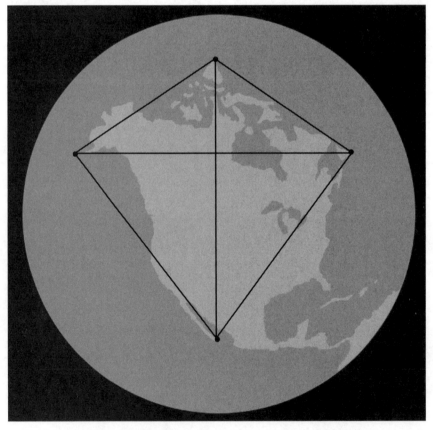

Figure 12.6. Four corners of the North American
continent forming a cross (shown as four right triangles).

west the point is on Unimak Island, just off the landmass limit of the Alaskan Peninsula.

The geometry between these North American extremes is once again coupled with phi digits. The two long diagonals measured from a point on the southern limits of the Baja California Peninsula are both 3,090 nautical miles in length and consequently have a total length of 6,180 nautical miles.

REPLICATION ON THE CONTINENTS

The northern part of the North American kite-shaped figure shows an equidistant relationship between Ward Hunt Island and the eastern and western extremes at Cape Spear and Unimak Island.

In North America the northern point is equidistant from these eastern and western points, a pattern that is replicated in a remarkable way on the South American continent. The eastern extreme of South America's landmass is Ponta do Seixas; the western extreme is found at the end of a thirty-kilometer-long straight beach at Punta Balcones. These two peninsulas are equidistant from the base of the northern peninsula of Greenland, Kap Morris Jesup. The duplication of equidistant relationships is illustrated by figures 12.7 and 12.8.

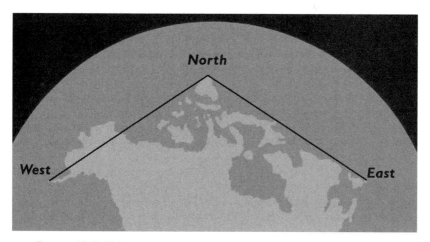

Figure 12.7. Western and eastern extremes of North America are equidistant from the northern extreme.

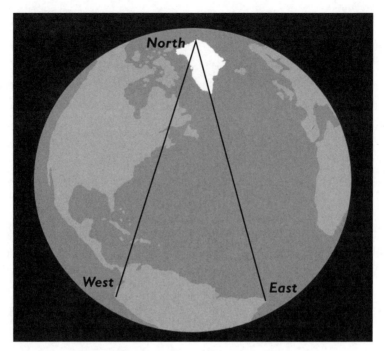

Figure 12.8. Western and eastern extremes
of South America are equidistant from the
northern extreme of Greenland.

The north point of Greenland is a long, curved icebound finger of land at the northern landmass limit. It is called Kap Morris Jesup, after the American philanthropist. With the exception of a couple of small rock islands and sandbanks, Kap Morris Jesup is not only Greenland's most northerly extreme, it is also the northernmost area of land on the Earth.

To create the isosceles relationship illustrated in figure 12.8, the geometric points on a map are located at the *base* of Kap Morris Jesup in the north, at the *base* of Punta Balcones's long beach in the west, and at the tip of Ponta do Seixas in the east.

The image does not fully illustrate how unusual the equidistant relatonship is and how it bears familiar hallmarks. The tip-to-base measurements found among the monuments are a replication of the same phenomena in nature. The apex in the geometry is at the base of Kap Morris Jesup, from where it is possible to measure *due south on the meridian* for exactly 5,400 nautical miles to reach the eastern extreme of the American landmass, at Ponta do Seixas. From the same point at the base of Kap Morris Jesup, 5,400 nautical miles can also be measured to the base of the Punta Balcones peninsula, at the western continental corner. The result is an isosceles triangle with one leg aligned to the meridian and with each leg measuring out one-quarter of the Earth's circumference. The two base angles of the triangle are, therefore, 90 degrees, although these angles do not appear as 90 degrees in figure 12.8 due to the curvature of the Earth.

Despite the unlikely probability of this isosceles geometry occuring repeatedly, the same equidistant relationship is found again across the Atlantic.

EUROPE AND AFRICA

Europe's northern landmass extreme at Cape Nordkinn is equidistant from the eastern and western extremes of Africa's landmass (figure 12.9).

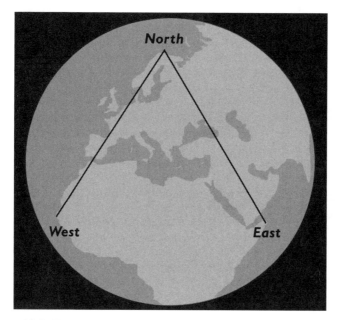

Figure 12.9. The tip of the Dakar Peninsula (West) and the tip of Dante Island (East) are equidistant from the base of Cape Nordkinn.

As with the apex in Greenland, the apex of this isosceles triangle in Europe is found at the base of the continent's most northerly peninsula.

The northern extremes of America, Greenland, and Europe all have an equidistant relationship with eastern and western continental extremes. This amounts to a pattern, and the sense that cohesive geometry is dictating the location of these continental extremes is echoed again when measuring from the northern extreme of Asia, at Komsomolets.

ASIA

Komsomolets is equidistant from Africa's eastern extreme, Dante Island, and India's southern extreme, Cape Comorin (see figure 12.10).

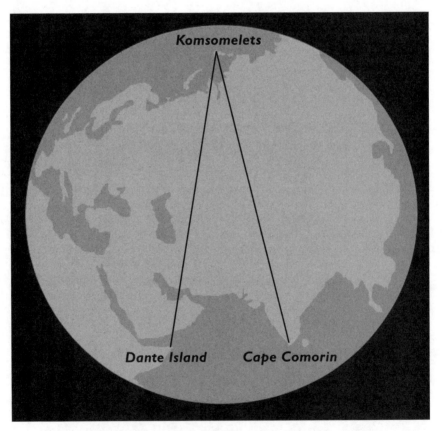

Figure 12.10. Dante Island and Cape Comorin are equidistant from a point on the southern reaches of Komsomolets.

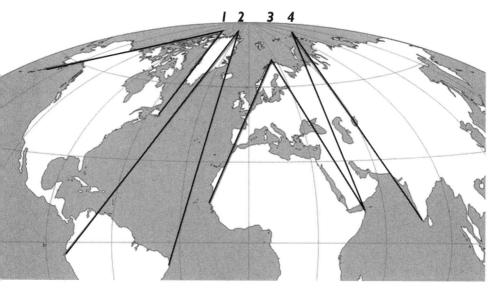

Figure 12.11. Four of the equidistant relationships between continental extremes and peninsulas, shown left to right: (1) Unimak Island, Ward Hunt Island, and Cape Spear; (2) Punta de Balcones, Kap Morris Jesup, and Ponta do Seixas; (3) the Dakar Peninsula, Cape Nordkinn, and Dante Island; (4) Dante Island, Komsomolets, and Cape Comorin.

The geometric point on Komsomolets is also equidistant from Africa's western extreme and Asia's southern extreme. Both apexes are found in an area about one kilometer in diameter on the island of Komsomolets. Four of the equidistant relationships are illustrated together in figure 12.11 (with considerable distortion owing to the curvature of the Earth).

The equidistant relationships shown in figure 12.11 are reminiscent of the geometry found in the United Kingdom in figures 6.2 to 6.6. On four occasions in the United Kingdom, the northern landmass points proved to be equidistant from landmass extremes to the south.

In almost any other context, this geometry would suggest that there is a systematic distribution of some description. But it is already well established and generally accepted that a pattern of this magnitude cannot have been fashioned; it must be accidental. Perhaps this is the fundamental difference between ancient and modern views of the Earth.

Figure 12.12. William Blake's *The Ancient of Days:*
measuring with dividers.

MR. EVEREST

Until his retirement in 1843 George Everest, whose name is pronounced "Eve-rest," was the Surveyor General and Superintendent of the Great Trigonometrical Survey of India. At the time the Great Arc of India remained incomplete, and the world's highest mountain was still unknown. But Everest, according to Sir Clements Markham, President of London's Royal Geographical Society, had been party to "one of the most stupendous works in the whole history of science." It was Everest's successor, Andrew Scott Waugh, who, after years of checking and rechecking copious quantities of data, concluded that Peak XV on his chart was the highest in the world. In 1856 he wrote, "To perpetuate the memory of that illustrious master of accurate geographical research, I have determined to name this noble peak of the Himalayas Mont Everest."[2] The title was shortly altered to Mount Everest and was endorsed by the Secretary of State for India and by the Royal Geographical Society. It seems the name has been mispronounced by the world ever since, though apparently not by the custodians of Everest's Himalayan home, now a museum.

While Mount Everest is the world's "highest mountain," the attribute of the world's "tallest mountain" goes to Mauna Kea in Hawaii. The flanks of Mauna Kea rise directly from the seafloor for 18,900 feet before breaking the surface. They continue to rise 13,796 feet above sea level, giving a total height of 32,696 feet directly from the seafloor.[3] This makes Mauna Kea notably taller than Mount Everest, which is a little over 29,000 feet. It was remarkable to find that 6,180 nautical miles can be measured from the tip of the world's "highest mountain" to the flank of the world's "tallest mountain,"* a measurement all the more interesting because it is the same distance that separates Mount Aconcagua from both Mount Kilimanjaro and Mount Elgon.

Both Mount Everest and Mauna Kea are five thousand nautical

---

*A line from the summit of Mount Everest 6,180 nautical miles in length extends to the flank of Mauna Kea above the water line.

miles from the southern reaches of Greenland. Consequently, an isosceles triangle can be drawn with a perimeter of 16,180 miles, with Mauna Kea and Everest at the base corners and with the apex located on the chevron-shaped tip of Greenland, as shown in figure 12.13.

We have come full circle, back to Greenland.

TIPS AND TOPS, ODDS AND ENDS

The visual association between chevron-shaped headlands and high points, epitomized in figure 12.13, could be seen as a symbolic language presented on the Earth itself (i.e., "signs on the earth"*). Did the monu-

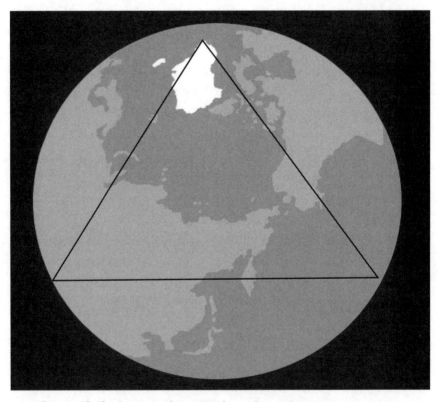

Figure 12.13. An isosceles triangle with a perimeter measuring 16,180 nautical miles. The corners are located on Mauna Kea, Mount Everest, and the tip of Greenland.

---

*Acts 2:19 English Standard Bible.

ment builders adopt their language of symbols by copying a language of symbols recognized in nature?

Apart from India, perhaps the world's most striking example of a chevron-shaped landmass is the Sinai Peninsula, and Egypt's highest summit, Mount Katherine, is located on it. Measuring from this summit to the very southern tip of the Sinai Peninsula, Google Earth gives a bearing of 161.8 degrees (figure 12.14).

A similar relationship is found on the Italian peninsula. If a line from Western Europe's highest summit, Mount Blanc, is extended over the tip of Italy's heel, it passes to the Giza plateau. The location of the Great Pyramid at Giza is therefore on a line drawn between an extreme point and a high point. The unique nature of all the alignments

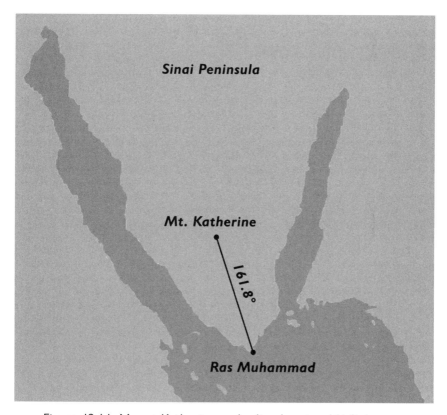

Figure 12.14. Mount Katherine and a line bearing 161.8 degrees from the mountain summit to the tip of the Sinai Peninsula, at the southern extreme point of Ras Muhammad.

intersecting at this Egyptian location is further emphasized by drawing a line from the central pyramid on the Giza plateau to the regional high point of Gebel Ataqa. This is the highest mountain overlooking the Nile Delta from the east and is located about 120 kilometers from Giza. Of all the possible bearings from Giza to Gebel Ataqa, it is extraordinary to find that a line bearing 90.00 degrees due east, taken from the summit of the central pyramid of Kaphre at Giza, extends directly to the summit of Gebel Ataqa with extreme accuracy.* The distance along this line is 61.80 + 3.142 nautical miles, give or take fifty meters. In other words, at the base of Kaphre's straight causeway, the Sphinx is gazing due east to the highest mountain summit directly overlooking the entire Nile Delta.

More than 3,500 years after Giza was completed, the pyramid-building phase of human history was drawing to a close, but it did not end quietly. There was a grand finale.

---

*The precise 90-degree bearing to Gebel Ataqa's summit on Google Earth is found from the center of Kaphre's pyramid. Using the Vincenty formula a line bearing 90 degrees from the center of Kaphre's pyramid passes within sixteen meters of the mountain summit.

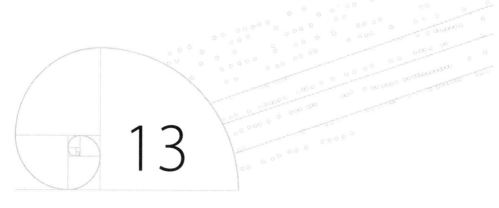

13

ANGKOR

ANGKOR WAT AND BOENG MEALEA

Among the vast complex of monuments at Angkor in Cambodia are the last of the great pyramids. About one thousand monuments were built at Angkor, from small shrines covering only a few square meters to the largest temple in the world, Angkor Wat. The Khmer King Suryavarman II, whose lavish coronation was recorded in 1119 CE, presided over the construction of the world-famous temple of Angkor Wat and (some say) another temple, Boeng Mealea, of similar size and splendor about forty kilometers to the east. If Angkor Wat really is the world's largest temple, then Boeng Mealea, although now desperately dilapidated, is perhaps the second largest. The location of Boeng Mealea is enigmatic. The great majority of the temples and pyramids of Angkor are clustered within a twenty-kilometer radius of Angkor Wat, but Boeng Mealea is well beyond this orbit. Despite the distance between the temples, the local road leaving the temple on the south side of Boeng Mealea is directed straight to Angkor Wat. Legend has it that the central peak (the *pyatthat*) of Angkor Wat symbolizes Mount Meru, the mythical mountain at the center of the Earth and home of the Hindu god Vishnu.*

---

*In addition to this use of the name for a fictional mountain, Mount Meru is also the name of one of the ten highest summits in Africa.

THE SUNSET PYRAMID

A city was first established in Angkor by Yasovarman I around 900 CE, and it was believed by some to be focused on a local high point called Phnom Bakheng, an isolated summit adjacent to Angkor Wat. (*Phnom* means "hill.") At the top of this hill a step pyramid, about 75 by 75 meters at the base, affords panoramic views to all distant horizons. Angkor is an extremely flat region punctuated by only a handful of strangely isolated natural hills, each summit offering expansive views of the whole area.

From the summit pyramid on Phnom Bakheng, the visitor scanning the eastern horizon cannot miss the two highest local summits of Phnom Bok and Phnom Dei standing up like enormous molehills on an unkempt lawn. Viewed from the pyramid these dominant summits line up with the eastern horizon of the Kulen Hills, as shown in figures 13.1 and 13.2.

Figure 13.1. Viewed from Phnom Bakheng's summit pyramid,
the flat summit of Phnom Bok aligns with the ridge of the
Kulen Hills in the distance. Photo by Mark Vidler.

Figure 13.2. Viewed from the same location as in figure 13.1, the hillside of Phnom Dei aligns with the falling escarpment of the Kulen Hills in the far distance. Photo by Mark Vidler.

The hilltop pyramid from where the pictures in figures 13.1 and 13.2 were taken is known locally as the Sunset Pyramid, due to the numbers of tourists who climb to the summit every evening for the spectacle. During a visit to the site in 2008, one corner of this structure was poised to collapse. Major repairs were underway, and hundreds of stones had been carefully removed and numbered prior to restoration. This work revealed the ancient foundation stones that had remained concealed since the pyramid's construction. These stones were perfectly cut together, one after another, along the full length of the foundation, 75 meters square . . .

"We cannot know what is underneath," said the engineer overseeing the work.

With these few words he described the situation at nearly every ancient monument in the world. We can never know who got there first.

The date for Phnom Bakheng's pyramid is established, but nobody can know what it conceals, or what, if anything, was removed prior to its construction. We cannot know who first stood on this hill, or who first noticed the alignment of adjacent hills, or who first revered this spot, or who left the first marker here. We can admire their efforts, the engineering skills, the craftsmanship, and the drive, but why was this hill, which is smaller than its two neighbors to the east, so distinguished?

SIMPLE LANDSCAPE GEOMETRY

Simple but beautiful hilltop geometry unfolds between these three high points overlooking Angkor. Figure 13.3 shows that points located on the summits of Phnom Bakheng, Phnom Bok, and Phnom Dei form an isosceles triangle aligned to the meridian.

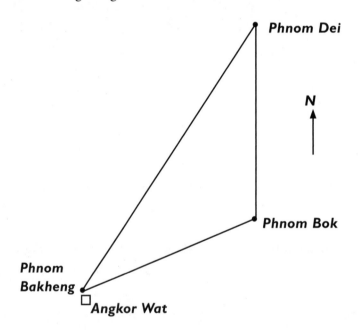

Figure 13.3. The isosceles triangle joining the natural regional high points overlooking Angkor. The summit in the north is Phnom Dei. Due south of Phnom Dei, at the apex of the triangle, is a temple on the summit of Phnom Bok. The Sunset Pyramid on the summit of Phnom Bakheng is located on the lower left, overlooking Angkor Wat.

If the Sunset Pyramid is considered as a single geometric point, then its position on the summit of Phnom Bakheng draws attention to the perfect isosceles triangle that can be drawn between points on the three summits. The alignment on the meridian is equally exact, within about 0.001 degrees of longitude, it seems, from summit to summit.* The GPS measurements for this area concur to within a few meters of those given by Google Earth.

With no known historical connection between Angkor and the European Neolithic, how could there be a common topographical understanding? Could the persistent refrain of "earth design," from Avebury to Carnac to Tiahuanaco, have somehow echoed through the millennia of monument building? The probability of three random points producing such an exact isosceles triangle, one that is all but perfectly aligned on a meridian, is extremely small. The ratio between the short side and the long side of the triangle shown in figure 13.3 is nearly 1:1.618, but not quite. In order to achieve the perfect phi ratio, the long side must be reduced by about 250 meters; the geometric point is then located at the base of Phnom Bakheng, at the beginning of the ancient straight causeway leading to the summit.

Was this natural geometry once recognized and perhaps revered at Angkor? If so, the Sunset Pyramid carries the same monumental message: one describing geometric order on the Earth.

THE GATES OF ANGKOR THOM

The Sunset Pyramid overlooks Angkor Wat and Angkor Thom, and both monument perimeters are within five hundred meters of the pyramid. Angkor Thom was once the "capital city" of Angkor. It is surrounded by a square moat with each side measuring about three kilometers in length. Enclosed within this is the centrally placed Bayon Temple, where more than two hundred enormous carved stone faces gaze to the cardinal

---

*Phnom Bok summit longitude, 103.983241 east. Phnom Dei summit longitude, 103.984274 east. Summit points established using elevations given on Google Earth.

points. Similar carved heads crown the arched gateways at the five entrances to Angkor Thom. Four of these entrances are centrally placed, one on each of the four walled sides of the square, but the fifth entrance is enigmatically located, off center on the eastern wall. Why was the symmetry broken, and how was the location of the fifth gate decided?

This "misplaced" fifth entrance is located on the line joining Phnom Bakheng and Phnom Dei. The line passes from the Sunset Pyramid, "in" through the southern gateway and "out" through the misplaced gateway on the east wall, terminating on the summit of Phnom Dei. In turn a line from the western gateway passes "out" through the misplaced gateway and continues to the summit of Phnom Bok (see figure 13.4).

The probability of one such chance alignment is small. The more distant probability of two alignments intercepting so precisely at one gate suggests that chance was not responsible for the location of the gates to Angkor Thom. They are aligned to the regional high points on purpose. These high points are therefore specified by the alignment process, and joining them together with lines, they form an isosceles

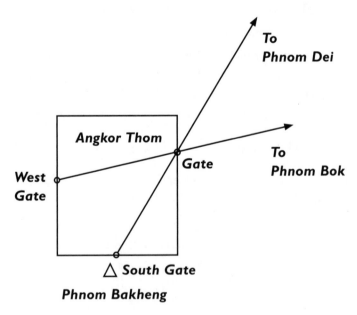

Figure 13.4. Illustration showing the alignment of the "misplaced" eastern gate at Angkor Thom.

triangle aligned to the meridian. The alignments at Angkor Thom, like those of Neolithic Europe, are therefore informative; they indicate, or specify, three geometrically ordered natural summits. (The Avebury alignments to four regional high points described in chapter 2 do almost exactly the same thing. Points located on the four specified hills can be joined together to create two isosceles triangles sharing one side.)

THE LOCATION OF ANGKOR WAT

The local geometry at Angkor extends to the broader topography. The alignment of a monument with a high point and an extreme landmass point is repeated at Angkor. Lines from the southern extreme of the landmass at Cau Mau in Vietnam can be drawn to Angkor Wat, to Angkor Thom, and to the dense cluster of pyramids and temples around them. All these lines pass over Cambodia's highest mountain, Phnom Aurul.

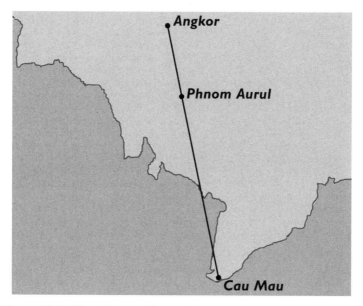

Figure 13.5. The cluster of temples and pyramids in the region of Angkor Wat and Angkor Thom all align with the body of Cambodia's highest mountain, Phnom Aurul, and with the southern landmass extreme at Cau Mau. The most precise alignment with the mountain summit is with the temple at Banteay Kdei.

The alignment of monument, high point, and landmass extreme has already become a familiar feature of monument locations in Neolithic Europe. The same alignment process continues in spectacular fashion at Angkor.

A line joining Boeng Mealea and Angkor Wat can be extended directly to the southern tip of India. The connection between cardinal extreme points and high mountains is followed through at Angkor because this line, from Angkor over the tip of India, continues to the summit of Mount Kilimanjaro, as shown in figure 13.6.

The linear language of symbols, ^-^-^, displayed at monuments elsewhere (e.g., Kilimanjaro, Carnac, and the southern tip of Greenland) is replicated on this line from Angkor. The symbols on the line are the highest summit on a continent and its neighboring summit, coupled with the distinctly pointed southern extreme of the Indian subcontinent, and two of the largest temples in the world.

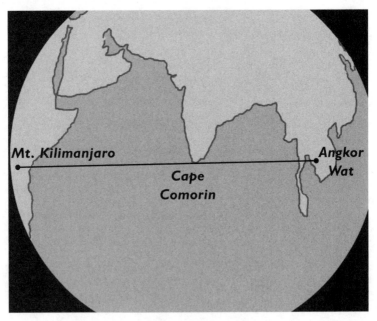

Figure 13.6. The line shown extends from the temple of Boeng Mealea, over Angkor Wat, over the southern landmass extreme of India, and on to the summit of Mount Kilimanjaro. From Kilimanjaro the line continues to the neighboring summit, Mount Meru.

This line passes along the road from Boeng Mealea, over one corner of Angkor Wat,* then over the breakers on India's southern landmass extreme point, Cape Comorin, to arrive on the crater's edge on the summit of Mount Kilimanjaro. It then continues to the summit crater of Mount Meru. The integration of Boeng Mealea and Angkor Wat into a natural alignment of topographical extreme points is, to all intents and purposes, perfect if the line is drawn on Google Earth. This is the first of two intercontinental lines that intersect at Angkor. The second joins the northern and southern extremes of the Asian continent.

THE ASIA MERIDIAN

At the southern extreme of Asia, Singapore Island is located between two southerly pointing fingers of land at the tips of the Malaysian Peninsula. The easternmost fingertip, near the humid and tropical beaches of Johor, shares the same meridian as the northern landmass extreme of the continent, the icy tip of Cape Chelyuskin. Boeng Mealea is located on this meridian, as shown by the line in figure 13.7.

In turn Angkor Wat is aligned between the tip of Cape Chelyuskin and the southern reaches of Singapore Island, also shown in figure 13.7. In other words two lines from Asia's northern landmass corner extended over Boeng Mealea and Angkor Wat and then continue to two of the continent's southern corners.

The second finger of land is found on the western tip of the peninsula, and this takes the form of a chevron. The tip of this chevron is the most southerly point on the contiguous landmass of Asia. From this extreme point a line to the causeway at the entrance of Angkor Wat has a bearing of 1.618 degrees; the line is more than 1,300 kilometers in length.

Both temples are aligned between the northern and southern limits of the Asian continent and draw attention to the fact that these two

---

*The line passes over the northeast corner of the moat.

Singapore Island and the
Southern tips of Asia

Figure 13.7. The southern extreme of Asia. The black line is the
meridian that continues northward to the northern landmass extreme
of Asia. This line passes over Boeng Mealea. The second line passes
over Angkor Wat and continues to the same northern extreme point.

continental extremes are located on the same degree of longitude.* This
north-to-south alignment of Asia's extremes, which passes unremarked
today, occurs against considerable odds of chance.†

Once again the Neolithic process of alignment with topographical extremes is repeated at Angkor. The code of practice requires the

---

*Longitude 104.25 degrees east passes over Cape Chelyuskin, then over the ancient res-
ervoir, or "Baray," at Boeng Mealea, and continues to the southeastern fingertip of Asia
at Kampung Tanjung Che Lahom. The southwestern "horn tip" at Tanjung Piai at 103.5
degrees of longitude is farther south.
†Drawing lines randomly over the continent very seldom achieves an alignment of two
cardinal landmass extreme points.

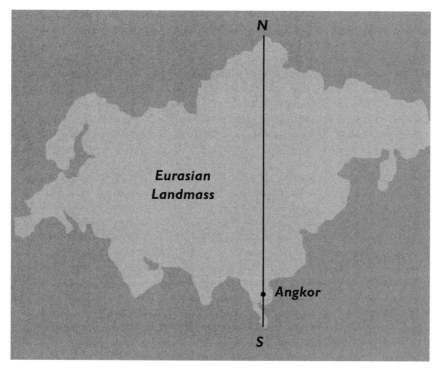

Figure 13.8. The location of Angkor aligned between
northern and southern extremes of Asia.

specified points to be joined independent of the monument that identifies them for us. This process invariably reveals some unusual topographical geometry as it does again here. A line joining the summit of Mount Kilimanjaro with the tip of Cape Chelyuskin extends to the tip of the Seward Peninsula. Thus, by following this process Mount Kilimanjaro is found to be aligned with two continental corners.

PHI ON EARTH

The specified points at Angkor include the southern extremes of India, Asia, and Vietnam. Geometric points located on each of these extremes can be joined together to form an isosceles triangle with legs that are 1,618 nautical miles in length.

By identifying, or specifying, these points the monument builders

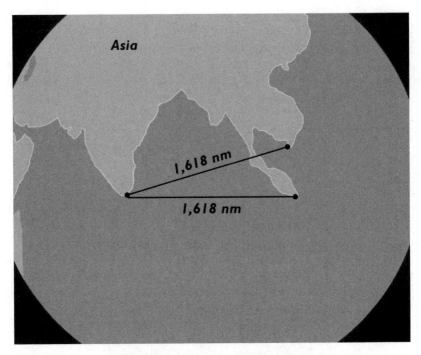

Figure 13.9. Lines drawn from Cape Comorin to
points located on Singapore Island and Cau Mau.

at Angkor illustrate the unusual relationship between these three southern continental extremes of Asia. But is this relationship really unusual, or is it part of a broader pattern?

The same distance separates the two southern peninsula tips of North America. From Isla Morada on the Florida Keys, it is 1,618 nautical miles to the southern extreme of the Baja California Peninsula in Mexico. There is a parallel between the two continents, as shown in figures 13.10a and 13.10b. The same distance can also be used to measure between the continental extremes in the north.

In figure 13.11 on page 164, the icy Asian limit of Komsomolets appears as a dot. From the same point, near the northern limit of Komsomolets, two lines extend to the limit of the Seward Peninsula and to the limit of Zenith Point. Both lines are 1,618 nautical miles long. Zenith Point is the northern extreme of the contiguous landmass of America; the Seward Peninsula is the western extreme.

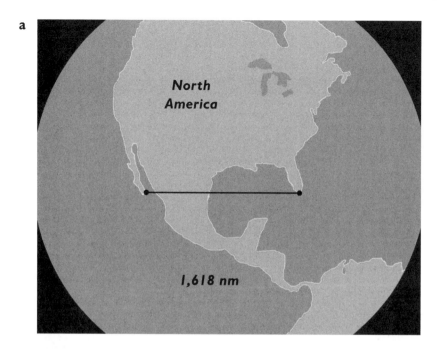

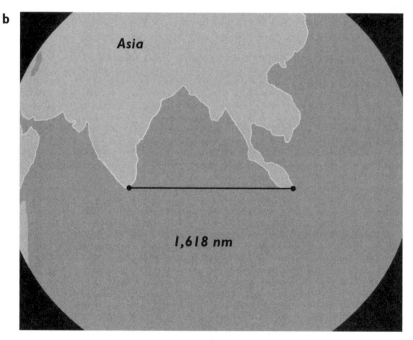

Figure 13.10. A distance of 1,618 nautical miles joins
extreme points in North America (a) and Asia (b).

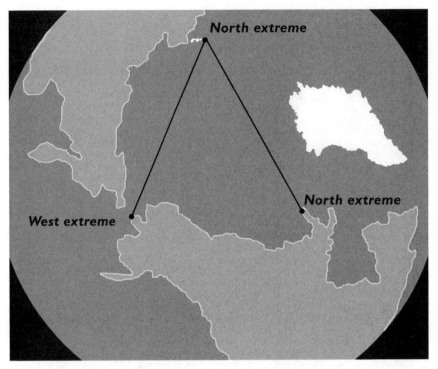

Figure 13.11. The northern and western extremes of
the American landmass are both 1,618 nautical miles
from a northern point on Komsomolets.

Something similar happens again on the other side of the globe, on the continent of Antarctica. The same distance separates the northern icebound extreme of Antarctica from the southern extreme, at the South Pole, as shown in figure 13.12.

And finally, at the top of the world, from the North Pole it is 1,618 nautical miles to the base of Mount McKinley.

The points specified by the monuments at Angkor are therefore part of a broader global pattern that exists between topographical extremes. This is exactly the same mysterious pattern defined by the builders at Carnac. There they used the same method of alignment adopted at Angkor, but they specified Mount Aconcagua and Mount Kilimanjaro, two extreme points 6,180 nautical miles apart. These pi- and phi-digit measurements are therefore part of a common code, or language.

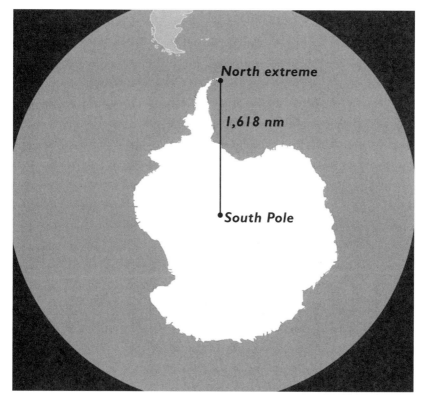

Figure 13.12. A line measuring 1,618 nautical miles from the South Pole to the ice islands on the peninsula tip of Antarctica.

The only language that can reveal the form and function of megalithic structures is that of number, the science of metrology (measurements based on the size and shape of the Earth), and alignment.[1]

Richard Heath, quoted above, is one among only a few people who currently recognize this language; nevertheless, it is in evidence as soon as the monuments are measured in the landscape.

At Angkor the northern and southern extremes of Asia are specified; at Avebury the eastern and western extremes of England are specified; at Loughcrew it is the eastern and western extremes of Ireland; at Carnac the northern and southern extremes of Africa are specified; at Cahokia the southern and western extremes of South America; at

Teotihuacán the northern extremes of both the Americas; at Almendres the eastern extreme of Africa and the western extreme of Asia; and at Giza the pyramids stand in the crosshairs between the eastern extremes of Africa and America and the southern and northern extremes of Africa and Asia. What a coincidence: all these monuments illustrate the same aspect of topographical geometry, some sort of elusive pattern in nature. By definition patterns have repetitive predictable elements. Can there really be a predictable pattern on the Earth?

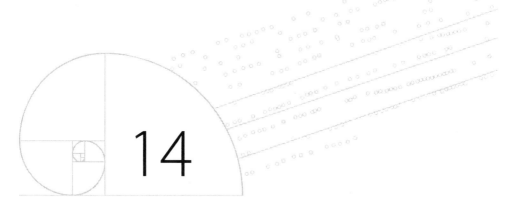

14

NEOLITHIC LANGUAGE

These hieroglyphics have evidently a meaning. If it is a purely arbitrary one, it may be impossible for us to solve it. If, on the other hand, it is systematic, I have no doubt that we shall get to the bottom of it.

SHERLOCK HOLMES, IN "THE ADVENTURE OF THE
DANCING MEN," BY SIR ARTHUR CONAN DOYLE

FATAL CODE

In the mysterious and dangerous world of spies and espionage, messages are frequently passed in code. In the past a cryptographer trying to understand a simple code commonly used a process called frequency analysis in an attempt to discover the hidden meaning. In the earliest and simplest ciphers, the frequency of symbols in the coded message reflected their frequency in the language itself, and it became possible to decode the message with this knowledge. Such was the case in Elizabethan England; the deciphering of Mary, Queen of Scots messages eventually resulted in her execution in 1587. Frequency analysis was used to decode a letter that allegedly provided "proof" of her plotting against the English Queen, which was sufficiently damning to convict her of treason. Thomas Phelippes deciphered the text using a process known as the frequency of common characteristics. Since Mary's misfortune, people passing secret information have become increasingly ingenious, yet even today faint echoes of the frequency analysis method can still be heard. The mind-numbingly complex methods of Alan Turing's Bombe, which deciphered German wartime messages encrypted by the Enigma machine, began with "a test." To make the test "the letter that appeared most frequently" was normally chosen by the cryptographers.[1]

Cryptography, like normal written language, depends on the ability to recognize symbols, but above and beyond this it generally depends on ordering the symbols on straight lines. The whole process can ultimately be reduced to ordering the lines themselves, and information can be transferred through the thickness and measured intervals between the lines, as is the case with the modern bar code.

The most popular and longest-prevailing methods of nonverbal communication are based on the simple alignment of symbols. Be it writing, music, or mathematics, the method is the same, and in all cases there is a secondary factor. The symbols have breaks between them, and without these measured intervals, the text becomes almost impossible to read—*Thereisnoeasywaytoreadlikethisallthetime.* The intervals between the symbols are fundamental to the communication process, and while this is obvious it illustrates how, at its root, communication at a distance depends not only on the symbols, but also on creating systematic distances between them. This is done automatically when writing, yet one seldom thinks of the measured intervals as part of the communication.

MEASURED INTERVALS

It becomes possible to "read" a Neolithic language by measuring the intervals between symbols on lines drawn over the landscape. And by measuring these intervals a natural and persistent pattern is revealed. On the small island of Mainland, north of Scotland, the Neolithic monument called Maeshowe encapsulates all the nuances and applications

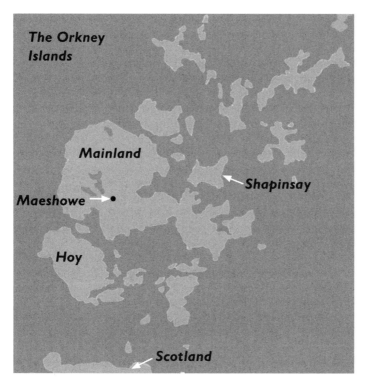

Figure 14.1. The Orkney Islands to the north of Scotland, showing
the location of Maeshowe on the island called Mainland.

of this ancient method of conserving and conveying information. It describes, if you like, the key to the code.*

There are three large Neolithic henge circles near Maeshowe on Mainland, and numerous standing stones and cairns are scattered among the fields and hillsides across the island. The settlement of Skara Brae on the west coast had long been known as the earliest example of relatively sophisticated Neolithic building in the United Kingdom; a complex of stone-walled rooms are knitted together, some still containing identifiable beds, alcove shelving, and internal drainage (circa 3000 BCE). The

---

*The key to the code is the sexagesimal system and the units of angle and distance derived by applying this base sixty system to the globe; that is, the unit distance is the nautical mile, the angle is in degrees. But just as we do today, the decimal system is adopted thereafter, resulting in 31.42 degrees and other pi- and phi-digit measurements (see chapter 4).

wonderful buildings of Skara Brae may have been hidden forever had it not been for a tremendous storm in 1850 that stripped the topsoil from a lonely coastal mound called Skerrabra and revealed the ancient dwellings beneath. But the complex at Skara Brae has been superseded by the discovery of a Neolithic complex located between two stone circles on the Ness of Brodgar, which is the subject of a June 2014 *National Geographic* article. A Neolithic building described by some archaeologists as a "temple" or "cathedral" has been unearthed on the ness, with walls four meters thick surrounded by a paved walkway. Nick Card, excavation director, commented, "The recent discovery of these stunning ruins is turning British prehistory on its head." He said, "What the Ness is telling us is that this was a much more integrated landscape than anyone ever suspected. . . . All these monuments are inextricably linked in some grand theme we can only guess at."[2]

A kilometer from this new archaeological site is the Ring of Brodgar, Britain's third-largest stone circle, once comprising sixty equally spaced stones in a true circle with a diameter of 103.65 meters. The Ring of Brodgar is at a focal point of lines, exactly as seen elsewhere in Neolithic Britain. In chapter 4 we found one line from the Ring of Brodgar follows a bearing of 6.18 degrees and passes to the northern landmass extreme of Mainland, Haafs Hellia. But a second line follows a bearing of 16.18 degrees and passes to the northern limits of Komsomolets.* These phi-digit measurements are now familiar due to the frequency of common characteristics found at one monument after another. But these are the present-day landmass limits and almost certainly differ from those at the time the monument was built.

Some seventy islands and skerries now make up Orkney, but ten thousand years ago the area was a single body of land. As the ice age came to an end, sea levels rose and the islands we see today gradually formed. Sea levels stabilized between 5000 and 3000 BCE, but have continued to alter by a few meters since the monuments on Orkney were

---

*This line is intercepted on the northern extreme of Komsomolets by a line bearing 1.618 from Mount Everest.

Figure 14.2. Maeshowe in Orkney: Europe's largest Neolithic
chambered cairn, and a geometric point in the landscape.
Photo by Malcolm Morris.

constructed.* A natural skepticism as to whether the current coastal
limits of Mainland could have been recognized in Neolithic times was
answered clearly when lines were drawn from the chambered cairn at
Maeshowe, about a kilometer from the Ring of Brodgar.

It was this investigation that led to the recognition of a predictable
topographical pattern that appears to extend around the world. The
mound at Maeshowe is located at a point where a hill *needs to be* in
order to complete a natural pervasive pattern.

THE MYSTERY OF MAESHOWE

Maeshowe, or the Mound in the Meadow, on the western portion of
Mainland, is about thirty-five meters across and seven meters high
and is surrounded by a circular earthwork about one hundred meters
in diameter. A narrow entrance corridor aligns to the highest point
in Orkney, Ward Hill on Hoy,† and leads to an interior chamber of

---

*Rising sea levels coupled with land that rose when released from the weight of the ice
created these islands.
†There are several Ward Hills. The highest summits on many of the islands of Orkney
are, rather confusingly, all called Ward Hill.

surprisingly spacious proportions. On the inner walls a range of graffiti bears testament to a group of Vikings who sought shelter there. To them the location of the monument was merely fortuitous, but why was this location chosen in the first place?

Figure 14.3 shows lines drawn from Maeshowe heading due north, due south, due east, and due west; these lines all pass to cardinal landmass limits.

The line traveling due east differs from the other three because it does not intercept a cardinal extreme on the western half of Mainland. However, it does pass to a notable headland, the Head of Holland, and then continues eastward, just missing a second more-chevron-shaped peninsula, Rerwick Head. These two headlands are indicated with arrows in figure 14.3.

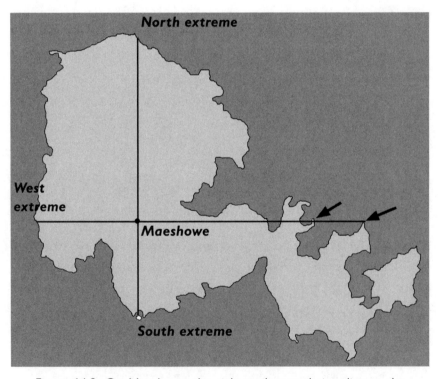

Figure 14.3. On Maeshowe the right-angle crosshairs align to the meridian, with lines joining the northern, western, and southern extremes of the western landmass of Mainland.

A close study of these cardinal alignments at Maeshowe revealed that they were apparently made intentionally to both the tip of a peninsula and to a point on its flank, or a base point. These alignments are described in appendix 2.

THE SIGNATURE MOTIF

The unusual thumblike peninsula of land, the Head of Holland, plays a special part in the geometry surrounding Maeshowe, as shown in figure 14.4. The use of the familiar digits of phi and pi in the topographical geometry around Maeshowe is coupled with a second familiar feature found among the circles in Cornwall, Stonehenge, Avebury, and elsewhere. This signature motif is an isosceles triangle and a right triangle sharing one side with one corner on the monument itself and with all the remaining corners located on topographical extreme points.

At Maeshowe this motif appears, but the two triangles do not share exactly the same line (see figure 14.5). The "shared side" of the triangles comprises two lines, one from Maeshowe to the eastern extreme of

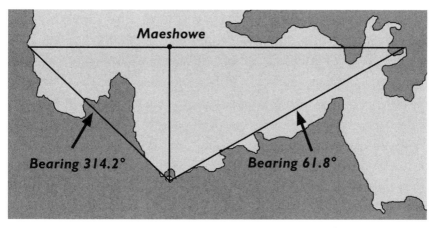

Figure 14.4. Two right triangles are drawn joining Maeshowe to the extreme points. One triangle hypotenuse has a bearing of 314.2 degrees, and the other, to the Head of Holland, has a bearing of 61.8 degrees.

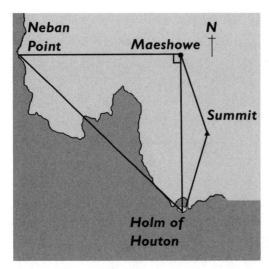

Figure 14.5. An isosceles and a right triangle drawn between Maeshowe and surrounding topographical extremes, including Ward Hill, the island's highest summit. The southern corners of the two triangles shown above are located at the eastern and southern tips of the Holm of Houton.

the Holm of Houton, and the other from Maeshowe to the southern extreme of the Holm of Houton (figure 14.5).

A MIRROR OF NATURE

The alignment of Maeshowe between cardinal landmass extremes is a direct parallel with the alignment of Avebury between England's eastern and western extremes. It is as if both monuments were *required* to be so aligned in order to maintain a rule or regulation. Applying this rule among the Orkney Islands, it becomes possible to anticipate aspects of natural topography. The key is given away at Maeshowe and appears clearly again at Avebury, Stonehenge, and other monuments. The underlying message describes a seemingly systematic alignment of high points with cardinal landmass extremes. The three largest islands surrounding Mainland illustrate this. The first is Hoy, the second largest of the islands in Orkney.

PREDICTABLE PATTERNS

Lines were drawn from the highest summit in Orkney, Ward Hill on Hoy, to the cardinal corners of that island. This was done in order to replicate the procedure at Maeshowe, but this time substituting the highest natural point for the man-made mound. The resulting series of alignments is shown in figure 14.6.

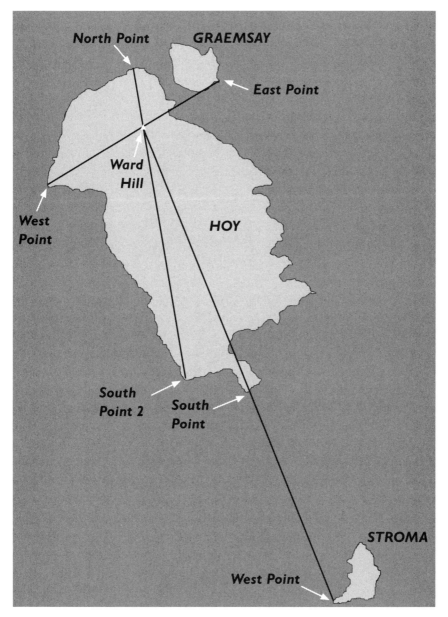

Figure 14.6. Lines drawn from Ward Hill, the highest point on the island of Hoy, over the cardinal landmass extremes.

The longest line shown in figure 14.6 joins Ward Hill and the southern limit of the landmass; the line extends to the western limit of the adjacent landmass of Stroma. Ward Hill is also found in the cross-hairs of lines joining northern, southern, eastern, and western landmass peninsulas, as shown in the figure. Each line passes within 150 meters of the summit point.*

Similar alignments were found on Mainland's two other large neighboring islands, Shapinsay and Rousay. On both these islands lines joining the two northern and the two southern extreme points intersect at the highest point on the landmass, as shown in figures 14.7 and 14.8.

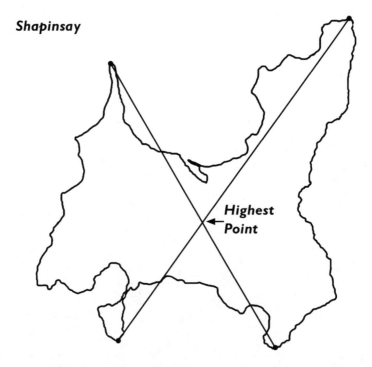

Shapinsay

Highest Point

Figure 14.7. Shapinsay, joining the northern and southern peninsulas. The lines intersect at the highest point on the landmass, Ward Hill.

---

*The summit of Ward Hill is relatively flat. The east-to-west line joins the tip of Rora Head in the west of Hoy with Sow Skerry on the tip of the Nevi on Graemsay. But the beaklike tip of Rora Head is 500 meters from the true western landmass extreme. Using the true western extreme, the line passes about 135 meters from the hill summit.

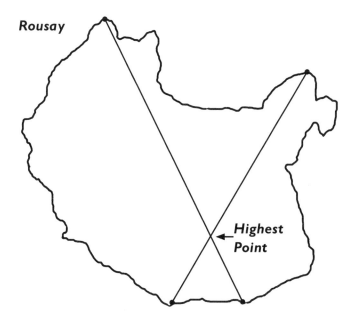

Figure 14.8. Rousay, joining the northern and southern extremes. The lines intersect at the highest point on the landmass, Blotchinie Fjold.

At both Rousay and Shapinsay, the geometric points are located on the coastal rocks at the cardinal limits of the landmass. On both islands the lines pass within 120 meters of the highest summit points. Landmass geometry such as this is tide dependent, and there are other variables, such as offshore skerries. For these reasons the geometric point itself has a radius. This radius (we believe) is defined at Maeshowe and at other monuments by the surrounding ditch or bank of a henge. The geometric points in the geometry are then found within this orbit, usually on the henge perimeter. More mysterious landscape geometry emerges on Shapinsay.

An isosceles triangle can be drawn between the two northern points on Shapinsay and the highest summit. The legs of this triangle measure 3.14 nautical miles. And then again an accurate isosceles triangle can be drawn between the two southern extremes of Shapinsay and the highest summit. The same applies with the southern extremes and the highest summit on Rousay.

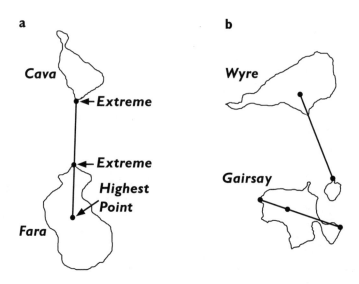

Figure 14.9. In both images an island's highest point aligns with two cardinal landmass extreme points. The same principle applies across Orkney.

The principle that applied repeatedly from one island to the next was simply that the highest point on the landmass will always align with two cardinal landmass extremes. These two extremes are either on the local island, or one is on the local island and one on a neighboring island. Examples of both these forms of alignment are shown in figures 14.9a and 14.9b.

To cut to the chase, high-point alignments like these were found on all the measurable islands in the group, some forty in all. The highest summit on the landmass is consistently aligned with two cardinal extremes on the local island or one on the local island and one on a neighbor. The only island where this was not true was Mainland itself. From this it could appear that the Mound in the Meadow was located where a high point should be in order to indicate that a natural pervasive pattern exists.

Maeshowe draws attention to this inexplicable yet natural phenomenon. An island's highest point aligns with island corners with a frequency that challenges current perceptions of an "accidental earth" and offers an alternative view. But this alternative view "stretches the mind beyond the bearable," requiring as it does "a ruthless metaphysics," as Georgio de Santillana and Hertha von Dechend wrote in describing the Neolithic period.[3] The challenges are not only to recognize what the Neolithic people apparently knew, but also to accept that they *could have known* it.

The islands that make up Orkney are so numerous that hilltop alignments with landmass extremes could perhaps be expected with regularity, given so many potential points. But as the island hopping extended around the world, the same rule of alignment applied virtually everywhere. One after another the highest point on an island is always aligned with two cardinal landmass extremes in the manner found in Orkney. The most disturbing thing about this is that the same cannot be said when random points are dotted around the islands. The lack of any consistency in the alignment of landmass extremes with random points distinguishes the natural topographical extremes from random points. The inevitable conclusion is that the island summit points are *not* distributed in a wholly random manner.

GOING TO EXTREMES

It should have been easy to find exceptions to the alignment principle, but a virtual dream cruise, traveling from island to island around the world, proved otherwise. From the Azores to the Canary Islands,* the Greek Islands, the Caribbean Islands, the Cape Verde Islands, the Solomon Islands, Indonesia, and the New Hebrides, fully 98 percent of the islands measured conform to the same consistent pattern. It is a near certainty that the highest point on any island in the world will align with two land-mass corners in the manner found in Orkney.† This pattern of alignments

---

*Mount Tiede on the Canary Island of Tenerife is one of the few exceptions found among more than one hundred islands studied. In each of these exceptional cases, two adjacent island high points align with a cardinal extreme point.

†Alignments are either to the tip or base of the cardinal landmass peninsulas. The two peninsulas are either local, or one is local and one is on an adjacent landmass.

appears to be "everywhere," exactly as Hermes says in the *Hermetica,* and it is this alignment principle that the monument builders repeatedly copy. Figures 14.10a–d and figure 14.11 provide a series of examples showing the repetitive nature of these alignments in nature.

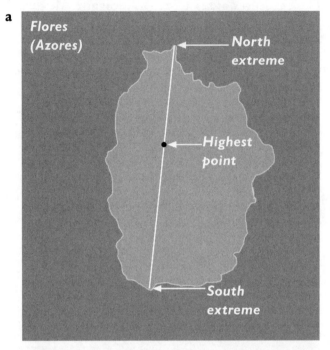

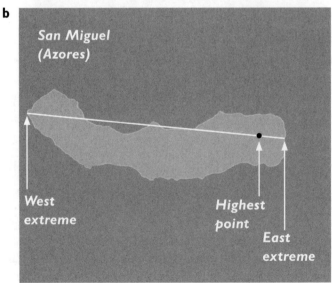

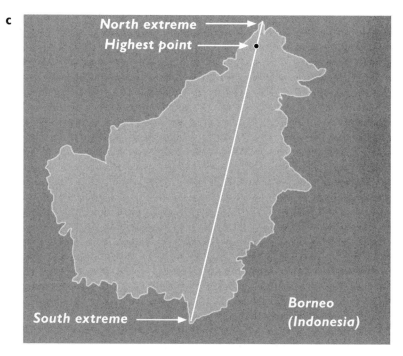

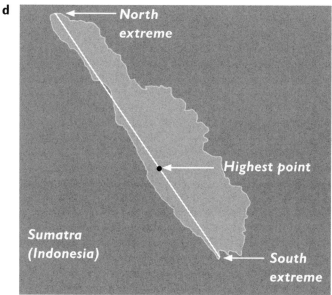

Figure 14.10 (a–d). The first two images, a and b, show the highest point aligned with cardinal landmass extremes on two Azores islands, Flores and San Miguel. The same principle applies on the much larger landmasses of Borneo and Sumatra, as shown in figures c and d.

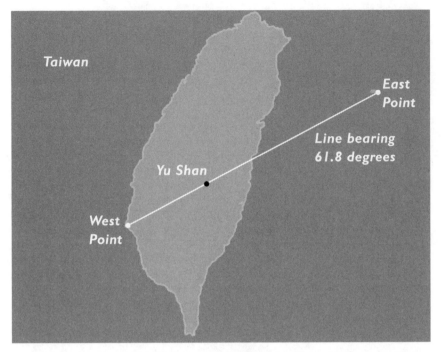

Figure 14.11. Taiwan and its highest summit, Yu Shan. The mountain is aligned with a chevron-shaped peninsula on Taiwan's western extreme and with the eastern extreme of the very small neighboring island, Yonaguni. This line has a bearing of 61.8 degrees.

Alignments such as those shown in figures 14.10a–d occur on about 33 percent of islands. But on 65 percent of occasions, the alignments are made with adjacent cardinal landmass extremes separated by water; examples are shown in figures 14.11 and 14.12.

Much of the western corner of Taiwan is reclaimed land. A line bearing 61.8 degrees begins on the western chevron-shaped peninsula, passes over the summit of Yu Shan, and continues to the eastern extreme peninsula of Yonaguni. Such alignments commonly define a point at the base of the cardinal peninsula, as is the case with the summit alignment of Mount Fuji, as shown in figure 14.12.

The line shown in figure 14.12 has a bearing of 31.42 degrees taken from Mount Fuji's summit crater. Having passed over the eastern extreme of the landmass, it continues to a distinct base point on the

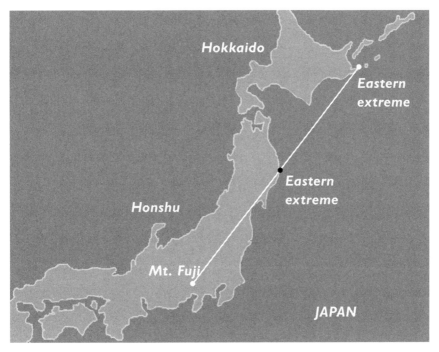

Figure 14.12. From a geometric point located on the summit crater of Mount Fuji, a line bearing 31.42 degrees extends to the eastern extreme of Honshu. The line continues to the eastern peninsula of the neighboring island of Hokkaido.

eastern extreme peninsula of the neighboring landmass. In this way such lines consistently "cut through" or "cut off" a cardinal extreme point on a landmass.

In figure 14.13 a line from Britain's highest summit, Ben Nevis, cuts off the northern extreme peninsula on the landmass, Dunnet Head, and continues to the southern extreme of the neighboring island of Hoy.

Exactly the same principle applies when a line is drawn from Ireland's highest summit, Carrantuohill. This line cuts through the northern peninsula of the Irish landmass and proceeds directly to the western extreme of mainland Britain near Ardnamurchan Point in Scotland, as shown in figure 14.14.

In this respect the topography on this island tour became predictable.

Figure 14.13. Mainland Britain's northern extreme cut off by a line joining Ben Nevis and the southern extreme of Hoy.

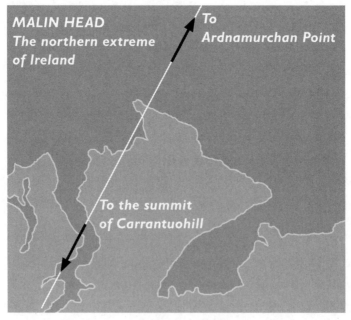

Figure 14.14. Mainland Ireland's northern extreme, cut off by a line from Carrantuohill to Ardnamurchan Point.

The same rule applied 98 percent of the time. Either the highest point aligns with two cardinal extremes of the local island (the island it sits on), or the line will join the highest summit with one cardinal extreme from the local island and one from a neighboring island. The precise summit alignment will frequently occur with a point at the base of one cardinal peninsula joined with a point at the tip of the other. In other words these lines locate the land area at the cardinal extreme, and when this is a peninsula, more often than not the line passes over the base of the peninsula, or if it is a headland, the line passes over the beach at the base of the headland. And again, when random points are located on the islands, they do not consistently adhere to this rule.

SRI LANKA

It was a struggle to find exceptions to this "island rule" until we reached Sri Lanka. The highest point on the island, Piduruthalagala, is not aligned between any two cardinal corners of the island, and there are no adjacent islands to which a satisfactory cardinal alignment can be drawn. Had something been missed?

There are some extremely small skerries isolated in the Indian Ocean about twelve kilometers off the west coast of Sri Lanka. These remote and isolated rocks are, rather eerily, precisely where they need to be to create an alignment with Piduruthalagala and the western extreme of Sri Lanka. A distance of 161.8 nautical miles can be measured from the skerries in the east to the peninsula in the west, although the precise western extreme on this peninsula is 161.3 nautical miles from the skerries. A bearing of 314.2 degrees can be taken from the mountain summit to the base of this peninsula; the line across Sri Lanka is shown in figure 14.15.

The alignment of landmass extremes held fast in Sri Lanka but illustrated that sometimes an offshore point is elusive.*

The same principle found among the smaller islands also applies to

---

*Fifty islands with this type of alignment are illustrated at www.neolithiconline.com.

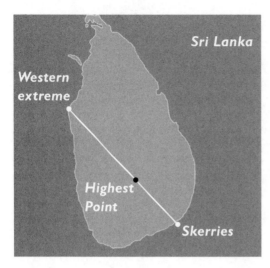

Figure 14.15. The line joining Sri Lanka's western extreme peninsula to the offshore skerries in the east. The line passes over the island's highest mountain summit. The offshore skerries span about three hundred meters.

the larger islands of the world. Such is the case on the world's largest island, Greenland, as shown in figure 14.16.

A straight line can be drawn from the base of the peninsula on Greenland's western extreme, over the summit of Gunnbjørn Fjeld, to reach the extreme western limit of Iceland.

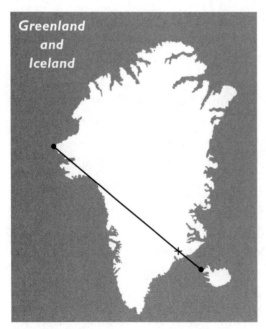

Figure 14.16. Greenland's highest mountain, Gunnbjørn Fjeld, is marked with a cross. The summit is aligned between the western peninsula of Iceland and the western peninsula of Greenland.

15

PI AND THE SONGLINES

SONGLINES

The creation myth of the native Australians, in which the first beings awoke from the land and "sang" its features into existence, is perpetuated in the Aboriginal concept of the "songlines." The name says it all: indigenous Australians are able to communicate the topographical features a traveler will encounter by singing them. Melodic phrases are used to indicate landmarks; singing these phrases for a particular line in the right sequence will give the traveler on that line a good idea of what lies ahead. And in a country peopled by innumerable tribes with different languages, dialects, traditions, and cultures, music—like mathematics—is a universal language and as such is the perfect medium.

Such a concept is not easy to understand. When Bruce Chatwin, in his semifictional travelogue, *The Songlines,* asks, "How the hell's it done?" he learns:

No one, he said, could be sure. There were people who argued for telepathy. Aboriginals themselves told stories of their song-men whizzing up and down the line in a trance. But there was another more astonishing possibility.

Regardless of the words, it seems the melodic contour of the song describes the nature of the land over which the song passes. . . . Certain phrases, certain combinations of musical notes, are thought to describe the action of the Ancestor's feet. . . . So a musical phrase . . . is a map reference?[1]

AUSTRALIAN CORNERS

In 1840, accompanied by two Aboriginal guides, Count Paul Strzelecki reached the highest summit on the Australian landmass. He named it Mount Kosciuszko because he believed it resembled a mound by that name in Krakow, built in honor of the Polish leader Tadeusz Kosciuszko.

The image of Australia shown in figure 15.1 has six lines on it; they join the four cardinal extremes of the landmass together. The landmass covers nearly eight million square kilometers, and looking at the picture it is clear how unlikely it should be to find the highest summit in Australia on one of these lines.

Figure 15.1. The image shows all six lines it is possible to draw joining Australia's four cardinal extreme points.

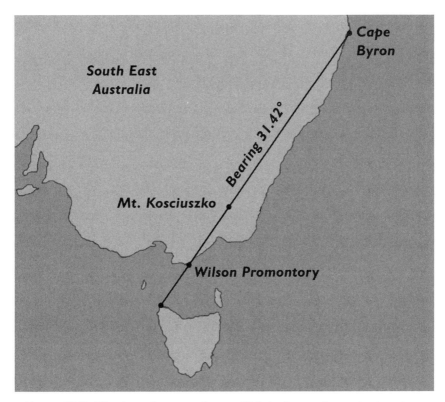

Figure 15.2. The line shown in figure 15.2 is drawn from the beach at Cape Byron, Australia's eastern extreme, over the summit of Mount Kosciuszko. It cuts through the southern landmass peninsula at Wilson Promontory and continues to Tasmania's northern extreme.

This enormous landmass follows the alignment principle recognized on smaller landmasses. Mount Kosciuszko is on the line joining the southern and eastern peninsulas of the Australian landmass, and the line continues to the northern extreme of Tasmania. The line from Mount Kosciuszko's summit to the beach at Cape Byron has a bearing of 31.42 degrees (figure 15.2).

The alignment of the Australian continent's highest point between two landmass corners occurs against considerable odds of chance, but it does no more than adhere to an alignment principle already recognized on the islands studied. And again, here in Australia, there is a pi-digit measurement between the mountain and the coastal extreme.

Looking at the two maps of Australia, the reader can judge the probability of finding this alignment and then couple this with finding pi digits when the line is measured. When the same improbable alignment and the same frequency of pi- and phi-digit measurements are found on almost all landmasses, the idea that the topography of these landmasses is entirely unregulated becomes increasingly dodgy.

If this alignment on the vast continent of Australia is compared with the alignment of Maeshowe on the small landmass of Mainland, it is possible to recognize the "echo" of the larger natural phenomenon made by the smaller man-made monument in Orkney. And likewise pi and phi digits are found when the topographical extremes are joined. The message from Maeshowe to Carnac describes an organized Earth, and the acceptance of such a possibility goes some way to explaining why monuments are located in alignments with topographical extremes. The ability of ancient people to grasp this as a global phenomenon may well be greater than mankind's today. Certainly, to us, it appeared both irrational and unbelievable.

So we continued to search for exceptions—with little success.

The alignment principle persists in Tasmania: the highest point, Mount Ossa, is aligned with the northern extreme of Tasmania at Robinns Island and the northern extreme of Hunter Island offshore.* A bearing of 161.8 degrees can be taken from Mount Ossa to Tasmania's southern extreme finger of land. This point forms one corner of an isosceles triangle with Mount Ossa and the island's eastern extreme.

NEW ZEALAND

Finding the unexpected is often what makes a journey of exploration exciting. But in our virtual topographical journey, the excitement came from finding what was by now to be anticipated. Is it now possible to

---

*Mount Ossa's summit aligns with the northern extreme of Hunter Island and the northern extreme of Robinns Island. Robinns Island is separated from mainland Tasmania by Robinns Passage, which is tidal and intermittently allows a land bridge between the two landmasses.

predict that the highest point in New Zealand will also be aligned with cardinal landmass corners?

There are two lines shown in figure 15.3, but they are too close to differentiate on this scale. The first line passes from the northern extreme point of North Island, and the second from the beach at the base of the northern point. Both lines extend to the southern extreme of South Island at Slope Point. The line from the beach in the north extends over the summit of Mount Cook, and the line from the extreme northern point passes over the flank of the mountain. In this and in the majority of cases we have studied, the precise line between the two landmass extremes passes over the flank or base of the highest mountain. In order for the line to pass over the summit, it will cut through the side

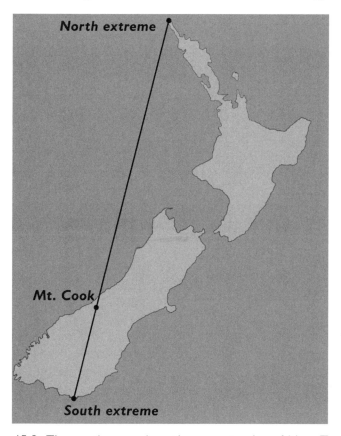

Figure 15.3. The northern and southern peninsulas of New Zealand align with the highest point in New Zealand, Mount Cook.

or base of the cardinal landmass limit (often at a beach or niche).* The constancy of this alignment principle in nature is remarkable.

Captain James Cook never saw the mountain named after him. Captain John Lort Stokes bestowed the honor because Cook was the first Westerner to circumnavigate the islands. To the indigenous Maori the summit is known as Aoraki and is considered sacred because it personifies one of their ancestors.

NORTH ISLAND

The highest summit on New Zealand's North Island is Mount Ruapehu. This mountain is also aligned between landmass corners, and once again the same tip-to-base or tip-to-flank alignment is evident.

Mount Ruapehu is one of the world's liveliest volcanoes; the first eruption is thought to have taken place about 250,000 years ago, and the most recent major eruption was in 1996. There are three summits surrounding the crater of Mount Ruapehu, and they have certainly not remained fixed throughout its long life. Yet even in this volatile landscape, the pattern is repeated on North Island because Mount Ruapehu falls into alignment with the eastern extreme of North Island and the northern extreme of South Island, with its long and distinctive spit of land called Farewell Spit.

Despite valiant efforts to save them, pilot whales tragically beach themselves on Farewell Spit with depressing regularity. About twenty-five kilometers from tip to base, the beaklike shape of this northern extreme bears a passing resemblance to a kiwi bird in profile. Figure 15.4 shows Farewell Spit with two lines that extend over Mount Ruapehu.

In figure 15.4 one line begins on the northern extreme point of the island's landmass and follows a bearing of 61.8 degrees to Mount Ruapehu. The second line also passes over Mount Ruapehu, but it continues to the eastern extreme tip of North Island, a total distance of 314.2 nautical miles; the extended lines are shown in figure 15.5.

---

*Such niches on the coast frequently have a right-angle appearance cut into the coast.

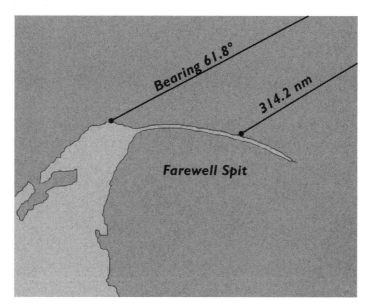

Figure 15.4. Farewell Spit, the northern extreme of South Island New Zealand.

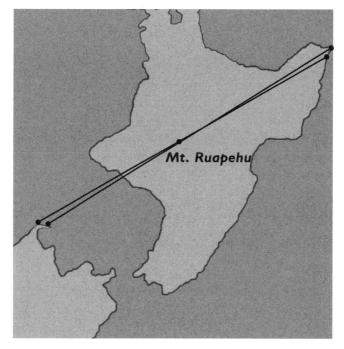

Figure 15.5. Lines from Farewell Spit to the eastern corner of New Zealand, passing over Mount Ruapehu.

These islands have not been specially selected to illustrate something uncommon. The highest summit aligns with local and adjacent cardinal landmass extreme points with a constancy not found with random points. Despite the considerable number of lines it is possible to draw between cardinal landmass extremes in New Zealand, the majority of points located at random on the landmass are not aligned between any two of them; consequently, the odds of chance are leaning against the high point appearing on one of these lines. The question is then, why is it so difficult to find islands where the highest summit does not appear on lines like these?

STEWART ISLAND

Just off New Zealand's south coast is Stewart Island. Figure 15.6 illustrates the principle of alignment once again, and once again there are pi digits in the measurement on the line.

There is something uncanny about the landmasses on Earth. For months new maps from around the world were studied. These were checked with various measuring tools, yet nothing altered the underlying pattern.

It is unrealistic to expect randomly located points in the landscape to consistently appear on lines joining landmass extremes, and to then anticipate pi- and phi-digit measurements associated with the lines. For this reason the realization that monuments and landmass summits more often than not conform to these criteria provides persuasive evidence that monument locations were decided *because the chosen point on the Earth offered a reflection of the same phenomenon occurring in the natural world.* This would explain why Avebury is located on a line joining England's eastern and western extremes and why this line passes over southern England's highest summit. The same natural high-point alignments are replicated on other islands. Avebury and other monuments around the world are therefore fitting into the predictable pattern. The meeting of two lines joining topographical extremes at Avebury only confirms that the location was carefully calculated, but the method

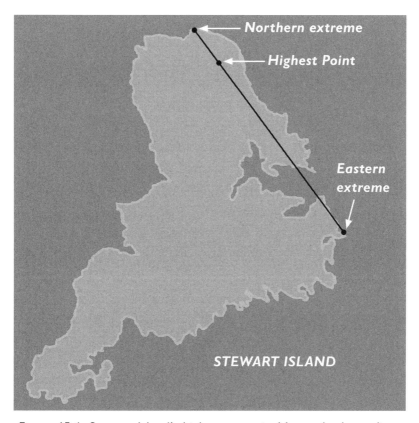

Figure 15.6. Stewart Island's highest summit, Mount Anglem, aligns between the eastern and northern peninsulas of the island. A bearing of 314.2 degrees can be followed from Mount Anglem to the base of the northern peninsula.

is unknown. Exactly the same applies at Giza, but the topographical extreme points in this case are continental landmass extremes. Both of these ancient sites feature manifestly topographical extreme points built by people, and both are located in the crosshairs of lines joining cardinal landmass limits. As more and more monument locations are added to the list—Stonehenge, the Ring of Brodgar, Carnac, and so on—it becomes increasingly clear that these monuments do not reside at randomly chosen points in the landscape.

Somehow, by some unknown method, seemingly preliterate people were capable of finding these locations. Whoever these people were their monuments convey an esoteric vision of the Earth for posterity. The pyramids of Giza and Teotihuacán are geometrically ordered man-made topographical extreme points located in alignment with naturally ordered topographical extreme points. The symbolism is explicit. Giza and Teotihuacán are aligned with continental corners because continental high points are aligned with continental corners.

Mount Kosciuszko in Australia has been shown to align with two continental corners. The same alignment with continental corners was seen with Mount McKinley* in North America and again with Mount Kilimanjaro in Africa.† In each of these three cases, the highest mountain *on the continent* is aligned with two or more cardinal *continental* extreme points. This effectively shows the same "island rule" working on a continental scale. So, do all the continent's highest summits conform to the same rule: Do they all align between continental corners?

GUAJIRA AND THE ALIGNMENT
OF ACONCAGUA

The northern extreme peninsula of South America, at Guajira, is no stranger to troubles. In 1891 most of the area was awarded to Colombia after a long-running border dispute with neighboring Venezuela, which was given only a thin strip of land on the eastern beaches. The Spanish explorer and cartographer Juan de la Cosa, who produced a *mappa mundi* in 1500, is popularly believed to be one of the first Europeans to set foot on the peninsula.

The Guajira Peninsula is shown in figure 15.7, with a line extending to the south. This line passes over the summit of Mount Aconcagua, the highest summit on both the North and South American continents.

---

*As shown in figure 11.8 in chapter 11. Mount McKinley is on a line joining the southern corner of Africa and the northern corners of America and Africa.

†"A line joining the summit of Mount Kilimanjaro with the tip of Cape Chelyuskin extends to the tip of the Seward Peninsula." See chapter 13. Mount Kilimanjaro is also aligned with the northern corner of Africa and the southern corner of Greenland. See chapter 8.

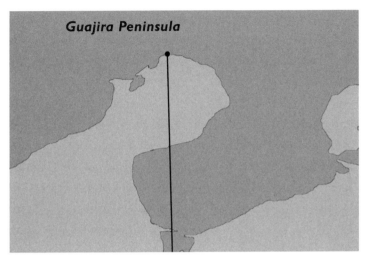

Figure 15.7. The northern extreme of the South American landmass, the Guajira Peninsula. The line from the northern extreme passes southward to Mount Aconcagua, the highest summit on the landmass.

Figure 15.8 shows the southern end of the same line cutting across a southerly pointing peninsula in South America. This peninsula is known as "False Cape Horn" or "Falso Cabo de Hornes" and it is found at the

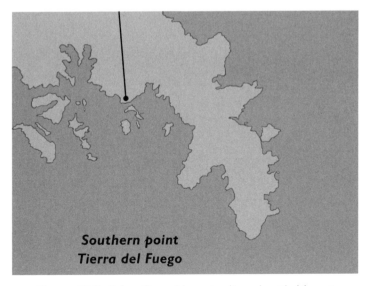

Figure 15.8. False Cape Horn is aligned with Mount Aconcagua and the northern tip of the Guajira Peninsula.

southern extreme of Horte Island, the southernmost point among the large islands of Tierra del Fuego. Excluding a cluster of offshore islands, where the "true" Cape Horn is found on Horn Island, this peninsula on Horte is the most southerly in South America. The peninsula is about forty kilometers in length, and the line from the northern extreme of the continent, passing over the summit of Mount Aconcagua, arrives, not at the tip, but at the base of this southern peninsula.

If the "False Cape Horn" is seen as an extreme continental point, the principle of the highest summit aligning with and cutting off (or through) the cardinal peninsulas applies to Mount Aconcagua in South America, just as it does to Mount Kosciuszko in Australia, Mount McKinley in North America, and Mount Kilimanjaro in Africa.

The same alignment principle is found again in Antarctica. The highest point, the Vinson Massif, aligns with the most southerly point on the landmass of Australia and the northern extreme tip of the Antarctic Peninsula. Viewed on Google Earth this line cuts through the very northern limit of the icebound peninsula, which is over twelve hundred kilometers in length. So for the fifth time the alignment principle holds true: each continental high point aligns with two continental extreme points.

Does the same principle apply in Asia?

THE ALIGNMENT OF EVEREST

A line more than sixteen thousand kilometers in length can be drawn from the summit of Mount Everest over the tip of Cape Comorin at the southern landmass extreme of the Indian subcontinent. The line continues, entirely over water, until it reaches land at Tierra del Fuego. As the line continues it reaches the contiguous landmass of South America about ten kilometers from Cape Froward, at the continent's southern landmass extreme. Therefore, give or take ten kilometers, by turning the world north to south, this line passes from the "apex" of South America's landmass, then over the "apex" of India's landmass, and on to the "apex" of the world's highest mountain, a massive ^-^-^.

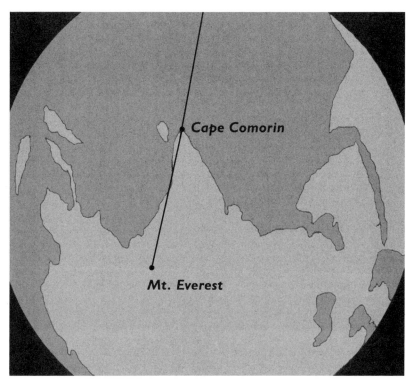

Figure 15.9. The line joining Mount Everest with two continental extremes, India's southern landmass extreme, and, out of image, South America's southern landmass extreme at Cape Froward.

The six continental high points all conform to the ^-^-^ principle.

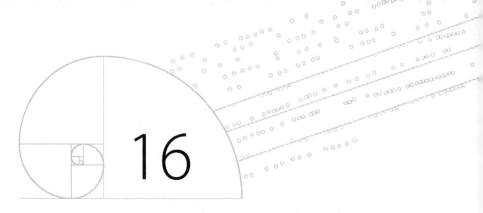

VIEWS FROM THE PAST

There are two distinct modes of scientific thought . . . one roughly adapted to that of perception and imagination: the other at a remove from it. It is as if the necessary connections which are the object of all science, Neolithic or modern, could be arrived at by two different routes, one very close to, and the other more remote from, sensible intuition.

LEVI-STRAUSS, *SAVAGE MIND*, 1966

IN THE MIND

The idea that an elusive advanced civilization once existed on the Earth has been suggested by numerous popular authors who have found their evidence in a variety of disciplines. To name but a few: Charles Hapgood saw evidence of this ancient intelligence in the form of maps; Paul Devereux in geography; John Michell, Alexander Thom, John Neal, Christopher Knight, Robert Lomas, Alan Butler, Robin Heath, and Richard Heath in metrology; Giorgio de Santillana and Hertha von Dechend in myths; Rand and Rose Flem-Ath in myths, geological catastrophes, and the legend of Atlantis; and Robert Bauval and

Graham Hancock in astronomical alignments. Fifty more authors could easily be added to this list. Does the diversity of this evidence reveal a sophisticated Neolithic mind, or is it all a grand illusion?

The monument builders convey an understanding of the Earth's topography, but their acquisition of this knowledge is paradoxical; solid evidence indicates that they didn't have the tools for the job. This absence of hardware for mapping the Earth sustains the orthodox view that no such mapping ever took place, or could ever have taken place, in antiquity. As we grasp for alternative explanations, the suggested scenarios include alien intervention. But this would suggest not one, but a series of interventions spread over thousands of years, occurring all over the planet. Or could the rash of monument building be attributed to some sort of *human* global consciousness that gave diverse peoples the same motivation? In other words did they collectively possess an understanding of the planet now lost to us? Was the apparently "alien intelligence" of the monument builders once our own, and if so, does it still exist today?

We cannot say how the monument builders chose their locations so accurately; there are numerous possibilities. It appears that the formal methods of measurement adopted today may not have been those adopted in the past, and if so, the solution to the problem may lie beyond the norms accepted in the present.

In the work of Guy Underwood, Hamish Miller, and numerous others, distinct characteristics are attributed to ancient monument sites. These characteristics are invisible. They are found through dowsing (often with twigs or metal rods) and consequently are frequently dismissed as nonsense. Underwood mapped out a spiral at Stanton Drew using only the pull on a twig for reference. Was Underwood's gift once more commonplace? If so, does the Earth itself identify these significant nodal locations in the landscape in a manner invisible to the majority today? The location of Stanton Drew, and other stone circles across the country, may have been identified without any long-distance surveys. We can only guess at the method adopted for locating monuments, but it could have involved human senses that are no longer generally acknowledged.

Another possibility is remote viewing.

REMOTE POSSIBILITIES

Between 1972 and 1995 the U.S. government funded a military program, at one time called Star Gate, which was specifically tasked to investigate the strange phenomenon called remote viewing. Information from diverse places is gathered by this process, but it does not involve the five known senses. It was dubbed remote viewing in 1971 by Ingo Swann, who first demonstrated this ability to scientists at the American Society for Psychical Research in New York. Swann was then invited to the Stanford Research Institute in California, when the Star Gate project began the following year; by the end of the project, more than 220 viewers had been involved.

An executive summary of the project, commissioned in 1995 and published by the American Institutes for Research, revealed that among selected data compiled over the preceding fourteen years, "a statistically significant effect has been observed" (i.e., an effect deemed highly unlikely to be caused by chance). The statistician, Jessica Utts, Ph.D., says in the report, "Using the standards applied to any other area of science, it is concluded that psychic functioning has been well established."[1]

The report is clear; it confirms that trained viewers, charged with the task of identifying so-called blind targets, were successfully describing the targets more often and more accurately than chance allows.

Why then does the study of human history allow no room for this possibility? Who can say if these abilities were once more commonplace?

This mass of scientific data now includes thousands of trials, and it has exposed a crucial aspect of human ability that has profound implications for the better understanding of our own history. Remote viewers today not only access remote locations, they also access the past, the present, and the future in exactly the same way as wise men, seers, soothsayers, and oracles reportedly did of old. In fact, it is hard to find any substantial mythical or religious text that does *not* include descriptions of remote viewing or something very similar to it. As well as visions and prophecies, other possible sources of remote viewing include dreams, revelations, being in the spirit, and even perhaps angels imparting wisdom. These ancient accounts are no longer considered to represent reality. We must believe instead that half the minds in the ancient world were bunged up with pseudoscientific imaginings, mystical claptrap, and magical hogwash.

ACADEMIC CENSURE

Regardless of the method adopted, the sophisticated, far-reaching terrestrial alignments at Stonehenge, Avebury, Stanton Drew, Rollright, Loughcrew, Carnac, Almendres, Angkor, Teotihuacán, Cahokia, Giza, and many more all illustrate the same thing. When viewed in a geometric context, the Earth's surface does not appear to be created by entirely random forces. This aspect of the landscape was somehow recognized at the time a monument's location was decided. John Michell and many others have already suggested something similar, but the inability to understand how this sophistication could possibly have occurred in prehistory has resulted in considerable academic cynicism and the total rejection of a copious quantity of legitimate and highly persuasive data. Remote viewing and dowsing are often likewise derided.

The people of the ancient world did not scribble the coordinates of Mount Everest on a piece of paper; they did far better. Their language is set firmly on the Earth. It is the language of measurement; their alphabet is numerical, and its meaning is still legible. When this understanding is applied to the lines at Nazca, the monument speaks for itself in a spectacular fashion.

THE CHEVRONS OF NAZCA

Be they henges, mounds, or pyramids, the majority of ancient monuments dramatically alter the landscape in which they sit, mimicking mountains and altering horizons as they point skyward. But at Nazca in Peru, this is not the case. The lines and geoglyphs drawn on the desert floor at Nazca cover about 450 square kilometers, but they barely alter

Figure 16.1. One of many chevron shapes
created on the ground at Nazca.

the topography at all. They are, quite literally, a scratch on the surface, created by shifting only a few inches of the topsoil to reveal the lighter earth beneath. There are no real high points on the Nazca plain; except for a few mounds a meter or two in height, the "monument" is essentially flat, or follows the natural contours.

The arid plateau at Nazca is one of the world's driest places, but it is covered with a vast confusion of long, intersecting straight lines that create an irregular, entangled mesh. The profusion of lines and geometric shapes was superimposed on scores of images: animals, birds, insects, and even a whale were all outlined on the desert floor at an earlier date. Interpretations of these huge, incongruous images have yielded an imaginative range of theories: they are astronomical, religious, or fertility symbols, to name a few. The current consensus among archaeologists is that the work on the plateau was carried out over about a thousand years, between 500 BCE and 500 CE.

REMOTE VIEWS

There are three clear indications at Nazca that some form of alternative viewing was achieved by the people who created this extraordinary land art. The first is the most fundamental: very few of the geoglyphs at Nazca can be successfully viewed from the ground. The artists dedicated untold hours of toil and sweat in order to create images that can only be seen with a remote view.

Second, they created depictions of animals and insects whose natural residences were also extremely remote, in some cases thousands of miles distant.

Third, as we will see, to understand the significance of the location of Nazca requires a worldview.

WHY CHOOSE THIS LOCATION?

There are innumerable long, straight lines at Nazca, but among the principle geometric figures on the plains are various chevron shapes, such as the one shown in figure 16.1. The purpose of these geometric shapes is unknown; as with other monument builders, their makers left no written account of themselves or their work.

Numerous measurements were taken from Nazca. It soon became clear that this monument did not fit into the pattern of alignments found in Neolithic Europe, Teotihuacán, Giza, and elsewhere. This monument repeatedly failed to direct us to the extreme points we expected. High points or landmass extremes, where would they lead us? Nowhere, it would seem.*

Eventually we realized that we had been looking not in the wrong place, but in the wrong way. Nazca is flat, so where would we find a flat extreme? The answer eventually came. The delta shapes at Nazca symbolize natural deltas, just as the pyramids and mounds symbolize mountains. With this insight the common code immediately became clear at Nazca.

1. A line 1,618 nautical miles long and bearing 61.8 degrees can be measured from Nazca to the Amazon Delta, the largest delta on the continent.

2. A second line 1,618 nautical miles long and bearing 31.42 degrees can be measured from Nazca to the Orinoco Delta, the second largest delta on the continent.

3. Extending the line bearing 61.8 degrees from Nazca over the Amazon Delta, it continues to the Nile Delta and onward for 3,142 nautical miles to the Ganges Delta.

The lines measuring 1,618 nautical miles extend from a geoglyph[†] at Nazca to the apexes of both the Amazon and Orinoco Deltas, at points where the rivers begin to fan out. Consequently, an isosceles triangle can be drawn between the Orinoco and Amazon Delta points,

---

*A single exception is Salcantay; the mountain can be reached on a bearing of 61.8 degrees from Nazca.
†The geoglyph is located at latitude 14.702775 s longitude 75.174461 w.

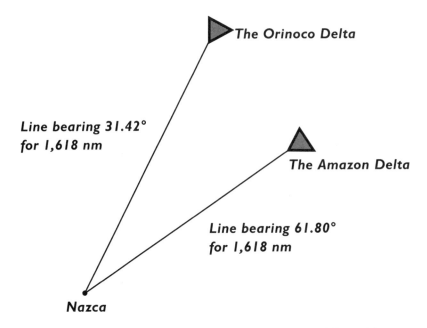

The Orinoco Delta

**Line bearing 31.42°
for 1,618 nm**

The Amazon Delta

**Line bearing 61.80°
for 1,618 nm**

Nazca

Figure 16.2. South America's largest deltas measured from Nazca.

with Nazca at the apex. The triangle can be drawn with legs measuring 1,618 nautical miles, and the bearings on the legs are 31.42 and 61.80 degrees, respectively.

Nazca is an extraordinary work of art at a phenomenal location on Earth. The delta-shaped geoglyphs at Nazca are on a line passing through three of the world's greatest deltas, and 3,142 nautical miles can be measured between two of them, the Nile and the Ganges Deltas. The line itself has a bearing of 61.8 degrees from Nazca.

The language is the same at Nazca as elsewhere, but this time ^-^-^ signifies three deltas in alignment with the ancient site, perhaps better represented by v-v-v. The deltas depicted on the ground at Nazca symbolize natural deltas, just as the pyramids, mounds, and henges symbolize natural high points.

The principle of crosshairs alignment, first noted at Avebury, is also seen clearly at Nazca. The two largest deltas on the North American continent are those of the MacKenzie River in Canada, which flows northwest and empties into the Beaufort Sea, and the Mississippi River

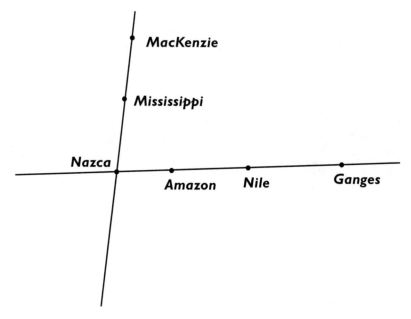

Figure 16.3. An illustration of the relationship of deltas with Nazca.

in the United States, which flows southward into the Gulf of Mexico. A line joining these two deltas intercepts the line joining the Ganges, Nile, and Amazon Deltas at Nazca.

Five of the world's greatest deltas are aligned with Nazca, and at the point where these natural "delta lines" converge, the desert floor is littered with flat, extended delta shapes etched into the Earth's surface. This raises more than a suspicion that the artists at Nazca knew where the world's largest deltas were, but because the origin of their understanding is elusive, orthodoxy will continue to demand that it never existed, that these lines of symbols, and the measurements between them, result from nothing more than coincidence.

At each of the sites visited in this book, the location of the monument has revealed a mysterious order in the natural world, such as the Kilimanjaro-Aconcagua-Elgon triangle with legs measuring 6,180 nautical miles or the Komsomolets–Cape Angela–Ward Hunt triangle with legs measuring 3,142 nautical miles. This geometry always results from joining together the points initially revealed by alignment with the monument. At Nazca these specified points are five large deltas. The common code of practice is tested by joining these points together independently. This procedure should reveal an esoteric understanding of the Earth; it should reveal an unrecognized geometric order.

THE DELTA CODE

When the deltas specified by alignment are joined independently, geometric order emerges. A line joining the Nile Delta to the MacKenzie Delta is at right angles to the line joining the Amazon, Nile, and Ganges Deltas, as shown in figure 16.4.

It seems that ancient monuments are all expressions of the same underlying belief, and at its root is the recognition of geometric order governing the topography of the Earth. The code at Nazca reveals that right triangles can be drawn between the Amazon, Nile, and MacKenzie Deltas; the Amazon, Nile, and Danube Deltas; the Ganges, Nile, and

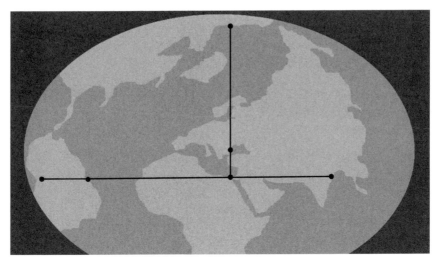

Figure 16.4. An illustration showing the right angle created by a line joining the MacKenzie and Nile Deltas. This line passes over the Danube Delta and is at right angles to the line joining the Amazon, Nile, and Ganges Deltas. The right angle is located in the central northern area of the Nile Delta. The horizontal line extends over three deltas to Nazca.

MacKenzie Deltas; and the Ganges, Nile, and Danube Deltas. A baseline shared by all these triangles extends directly to Nazca, where the bearing on the line is 61.80 degrees. It is a spectacular example of landscape art and earth science rolled into one.

MISSING EVIDENCE

The evidence of sophistication in the past is not missing. The deltas are aligned with Nazca, and the deltas are aligned with each other. Also, 3,142 nautical miles can be measured between the Nile and the Ganges Deltas, and then 1.5 × 3,142 nautical miles can be measured between the Nile and the MacKenzie Deltas. The overall message is again topographical; it describes the Earth's surface as something with the appearance of geometric calculation. Like the shell of a turtle, there is a regulated pattern. Whatever causes this pattern monument builders the world over were preoccupied with it. The ancient world appears to have reached a conclusion about this. The message passed down from early history is that of an ordered, fashioned, designed, or created Earth. The cause of this was variously attributed to living forces, supernatural forces, or a single god—all apparently deemed reasonable conclusions at the time.

When a line joining the Ganges, Nile, and Amazon Deltas is drawn around the world, the probability that the specific location of Nazca was chosen, on this line, without knowledge of the line, is palpably remote. Add to this the crosshairs with a line joining the Mississippi and MacKenzie Deltas, and there is simply no other location on the planet where Nazca *could* be to achieve this distinction. Consequently, to find uniquely formed delta shapes carefully crafted on the ground at this specific location on the planet and pass it off as a coincidence seems unrealistic, not least because of the common numerical motif of pi and phi digits found when measuring the lines.

Throughout this research a common geometric motif has also been found in the form of a right triangle and isosceles triangle sharing one side (many more examples of this are given in appendix 1). At Nazca

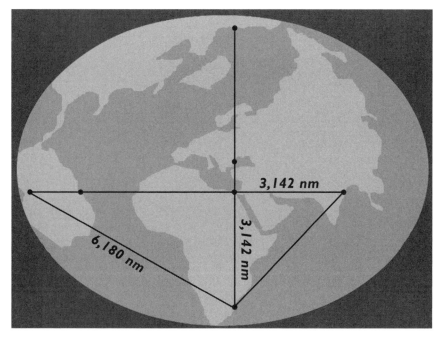

Figure 16.5. Measured intervals between aligned delta points and Nazca, with lines forming a right triangle and an isosceles triangle that share one side.

this motif reappears, and when measuring it, there is a remarkable confirmation that the code has been correctly followed.

Extending the line from the MacKenzie Delta, over the Danube Delta, and onward over the Nile Delta, the line continues southward to the delta of the Save River on Africa's southeast coast. Four deltas therefore rest on this single line, and the following measurements can be taken:

3,142 nautical miles can be measured from the Nile Delta to the Save Delta.

6,180 nautical miles can be measured from the Save Delta to Nazca.

Is it realistic to regard the location of Nazca in this geometry as "nothing more than a coincidence"? Figure 16.6 illustrates the full geometry between Nazca and the deltas.

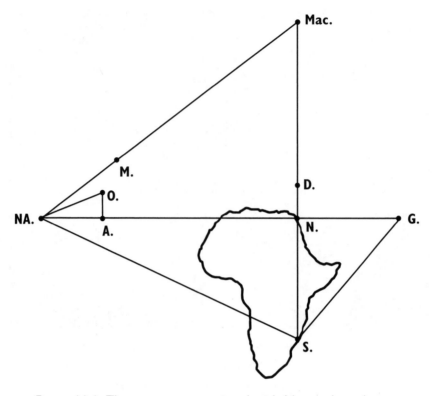

Figure 16.6. The geometry associated with Nazca. A. = Amazon Delta, D. = Danube Delta, G. = Ganges Delta, M. = Mississippi Delta, Mac. = MacKenzie Delta, N. = Nile Delta, NA. = Nazca, O. = Orinoco Delta, S. = Save Delta.

Summary of measurements that can be taken between deltas:

NA.–O. = 1,618 nautical miles
NA.–A. = 1,618 nautical miles
N.–S. = 3,142 nautical miles
N.–G. = 3,142 nautical miles
Mac.–N. = 3,142 × 1.5 nautical miles
NA.–S. = 6,180 nautical miles
NA.–O.: bearing 31.42 degrees
NA.–A.–N.–G.: bearing 61.80 degrees
Lines intersect at right angles at N

All at once the seemingly inauspicious location for a monument is recognized as a geometric point. These ancient monuments speak to the modern world and illustrate an alternative way of seeing our planet. They ask:

Why are these deltas so well ordered?
Why are the continental corners so well aligned?
Why are high points aligned with extreme points?
Why are there so many pi and phi digits?

LANGUAGE OF THE PAST

The many thousands of artifacts unearthed at archaeological sites around the world present a linear picture of human history, a picture of gradual progress from humble beginnings to the sophistication of the present day. This understanding has become the bedrock for current views of past cultures, and consequently, information that contradicts this understanding challenges the foundation of a commonly accepted principle. Occasionally, antique objects are discovered that appear to be out of place in the context of the historical period, but these are seldom seen as more than anomalies.* Likewise, the application of seemingly highly advanced mathematics and astronomy at very ancient sites is seldom recognized as confirmation that such advances had been achieved. The late Professor Alexander Thom fell victim to this intransigence; he presented clear evidence that Pythagorean geometry was in use at Avebury, long before Pythagoras was born. Thom's evidence demanded a conceptual shift because it revealed the use of technically advanced geometry, astronomy, and surveying in the Neolithic.

---

*These objects are called ooparts, meaning "out-of-place artifacts." The Antikythera Mechanism is the most celebrated example. This is a very precise mechanism with at least thirty gear wheels creating a replication of the movements of celestial bodies with near-perfect accuracy. The mechanism, dating from 100 BC, is about 1,500 years ahead of its time. Neolithic stone spheres depicting the five Platonic solids are among the more enigmatic ooparts.

A leading authority on the Neolithic at the time, the archaeologist Richard Atkinson, advised his colleagues that Thom's work needed to be taken seriously "because if we reject it we are flying in the face of a great deal of evidence which cannot be explained away simply as the result of chance or accident."[2] But it was too much for the establishment of the day; the essential change of viewpoint never took place, clashing as it did with an impenetrable orthodoxy.

A few faults were found in Thom's forty-year study of Neolithic geometry and astronomy, and for these he was "airbrushed out of history," as his biographer, Robin Heath, describes it.

Are students of prehistory today entirely misled by an institutionalized dogma? It is not simply Thom's work that evaporates in the lecture hall; the raison d'etre for monument building itself also disappears with it. In consequence the illusion that these immutable monumental masterpieces were built for some Neolithic numbskulls to dance around while wailing at the sunrise has become a standard history text, and this image percolates via the media to the public. This, for example, is the standard Neolithic scene at Stonehenge conjured by Aubrey Burl: "There would have been shouting, wails and screams, the blowing of horns, an excitement of noise . . ."[3]

This is supposition; there could equally have been a tradition of silence. But the paradigm of the flailing, wailing, screaming barbarian at Stonehenge, and elsewhere in early history, has been so firmly set that the genius required to construct the monuments in the first place is overlooked and ultimately becomes anathema.

The massive sarsen stones at Stonehenge are many miles from their original location. Why not do all that shouting, wailing, and screaming in the beautiful valley where the stones belong? The question as to why a particular location was considered significant has remained unanswered by archaeologists because it cannot be answered without recognizing the significance of a monument's position in the landscape. When this is done, be they stone rows, circles, or mounds, the monuments stand united in their common purpose: they are signposts in our landscape.

The wonderful geometric patterns in the petals of a flower, the veins of a leaf, and the shell of a snail are easily appreciated today, but the monument builders' vision was greater than this. They saw geometric order on the planet itself. Their enduring monuments convey this in a language of symbols with a common message:

It really is a wonderful world.

APPENDIX I

COMMON GEOMETRIC SHAPES AMONG EXTREME POINTS SURROUNDING CORNISH STONE CIRCLES

The following study of extant stone circles in West Cornwall includes the great majority of them; none have been excluded intentionally. The five most westerly circles are uniformly distributed in the sense that each location defines the same series of coastal extreme points by measurement. This is done by repeating the same geometric motif of a joined isosceles and a right triangle. This motif appears when lines are drawn from the circles to the cardinal landmass extremes on England's western extreme. The same motif is evident with the majority of circles on Bodmin Moor (fourteen are discussed below). When the lines creating this geometric motif are measured, the common unit distance is the nautical mile, 1,852 meters.*

---

*All the measurements given in this appendix can be found using the Vincenty formula. These distances concur with those on Google Earth, usually within ten meters. All measurements are taken from the area defined by the circles. Coastal extreme points are generally found within the current high- and low-tide limits at the cardinal landmass limit.

THE MERRY MAIDENS

The Merry Maidens stone circle is located near Land's End, England's most westerly point. The circle has a diameter of about twenty-five meters.

It is 5.00 nautical miles from the Merry Maidens circle to the tip of the Land's End Peninsula.* It is 6.00 nautical miles from the circle to the tip of Cape Cornwall, the second westerly extreme point on the landmass.

A right triangle can be drawn between the tip of Cape Cornwall, the base of the Land's End Peninsula, and the Merry Maidens circle.

The northern coastal extreme of this region is Clodgy Point, which is 10.00 nautical miles from the Merry Maidens circle. A tip-to-tip distance of 10.00 nautical miles can also be measured from Clodgy Point to Cape Cornwall; consequently an isosceles triangle can be drawn with these two points and the stone circle. The joined right triangle and isosceles triangle are shown in figure A1.1.

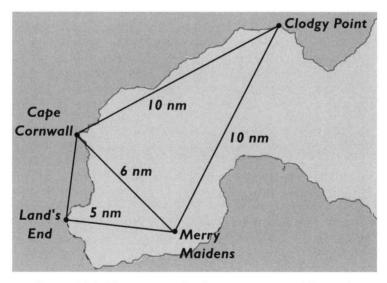

Figure A1.1. The western landmass extreme of England. The joined isosceles and right triangles are created between the landmass extremes at Clodgy Point, Cape Cornwall, Land's End, and the Merry Maidens circle.

---

*This peninsula tip consists of a single lump of rock about thirty by fifty meters.

The measurements are made with today's nautical mile value, which is 1,852 meters. The 5.00, 6.00, and 10.00 nautical-mile distances can all be measured from within the Merry Maidens to the rocks at the limits on each headland.*

The distance from the Merry Maidens to England's southern landmass extreme at Lizard Point is 16.18 nautical miles, within a few meters of the current southern landmass limit at low tide.

The geometry and the measurements indicate that the Merry Maidens was not a randomly chosen point. The measurements, in nautical miles, demonstrate awareness of this distance and of the current landmass limits in the context of geometric points. This awareness is common to all the circle builders. The use of the nautical mile and specific bearings can be anticipated with regard to the great majority of stone circles in southwest England. The circle builders' recognition of these coastal extremes becomes apparent through their repetitious references to them as geometric points.

TIP TO BASE

At first it appears coincidental, but one circle after another exhibits a subtle feature whereby the distance to a measured point and the location of the geometric point differ by a fractional amount. The difference is measured out between the tip of a peninsula (or a skerry off the peninsula tip) and a point on the peninsula's base. At the Merry Maidens the line to the tip of the Land's End Peninsula (shown in white in figure A1.2) is 5.00 nautical miles long on a bearing 3.14 degrees north of west. But when a point at the base of the peninsula is adopted (where the black lines converge in figure A1.2), a right triangle is formed with the Merry Maidens circle and the tip of Cape Cornwall. The difference between the points on the peninsula shown in figure A1.2 is about 250 meters.

The definition of tip and base points is a common factor when

---

*Measurements made using the Vincenty formula.

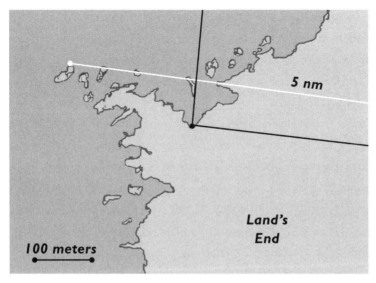

Figure A1.2. The Land's End Peninsula with the two points defining the tip and the base of the peninsula.

other stone circles are investigated. The dot at the end of the white line shown in figure A1.2 is 5.00 nautical miles from the Merry Maidens circle, and it is also 4.00 nautical miles from the Boscawen-Un stone circle. The same point is 4.50 nautical miles from the Tregeseal East stone circle. These are the three closest circles to Land's End. From the same point it is 12.50 nautical miles to the tip of Clodgy Point. The Merry Maidens identifies this odd numerical relationship. The two western extremes of Cornwall are 10.00 and 12.50 nautical miles from the northern extreme of the peninsula at Clodgy Point.

The Boscawen-Un circle is less than two kilometers from the Merry Maidens. By drawing lines from this circle to the same surrounding topographical extreme points, the common use of these present day extremes as geometric points becomes evident.

BOSCAWEN-UN

The landmass on the far west of Cornwall has its southern extreme at Hella Point, 3.92 nautical miles from the Boscawen-Un circle. The same distance can be measured from Boscawen-Un to the base of the Land's End Peninsula. An isosceles triangle can therefore be drawn between the circle, a peninsula tip, and a peninsula base. Extending the line to the tip of the Land's End Peninsula, the distance from Boscawen-Un is 4.00 nautical miles.

There is a single skerry about fifty meters off the tip of Hella Point; this skerry is the most southerly point in this region. The bearing from this rock to the Boscawen-Un circle is 31.42 degrees. Following this bearing to the north leads to the base of the northern extreme peninsula at Clodgy Point. However, if the same bearing is taken from the landmass extreme at Hella Point, the line extends to the northern tip of the peninsula at Clodgy Point.

In this way the Boscawen-Un circle builders specify the southern extreme skerry with a pi-digit measurement (just as the tip of Land's End is specified from the Merry Maidens), and by moving some fifty meters from the skerry to the landmass extreme point, the pi digits are found in nature, on a line joining the two landmass extreme points.

If further measurements are taken between these natural topographical extremes, a series of pi and phi digits emerge. A bearing of 61.8 degrees can be taken from the skerry at Hella Point to the skerries off the eastern landmass extreme at Penlee Point. In turn a bearing of 314.2 degrees can be taken from the skerry at Hella Point to the skerries off the Land's End Peninsula. In addition a bearing of 345.67 degrees can be taken from the skerry at Hella Point to the tip of Cape Cornwall, and this line forms the hypotenuse of a right triangle with the Boscawen-Un stone circle at the right-angle corner. There can be a sense of play associated with this ancient geometry. The joined right and isosceles triangles with a shared corner at the Boscawen-Un stone circle are shown in figure A1.3, with the line bearing 31.42 degrees joining the northern and southern landmass extremes. Owing to the scale of the illustration, the two lines bearing 31.42 degrees discussed above appear as one line.

Consequently, both Boscawen-Un and the Merry Maidens are locations that produce joined right and isosceles triangles when lines

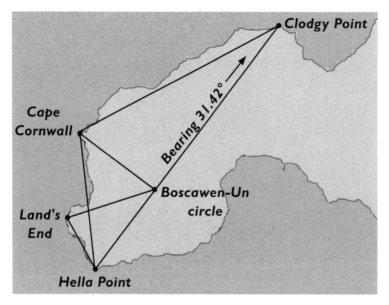

Figure A1.3. The isosceles and right triangles drawn
between Cape Cornwall, Land's End, Hella Point, and the
Boscawen-Un circle. The line joining the northern and
southern landmass extremes has a bearing of 31.42 degrees.

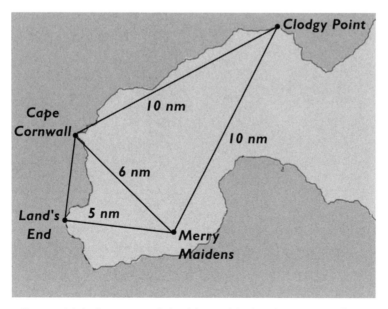

Figure A1.4. A repeat of the Merry Maidens' geometry for
comparison with the Boscawen-Un geometry.

are drawn to landmass extreme points, and both produce pi digits in bearings. Above and beyond this duplication, the result of studying the Boscawen-Un circle is to discover a pi-digit bearing between two natural northern and southern extreme points.

THE NINE MAIDENS (BOSKEDNAN)

The highest point overlooking this region at the western limits of England is called Carn Brae, and the contours on the Ordnance Survey map give the hill a roughly circular shape, which is evident when driving around it. A line bearing 31.42 degrees from the summit of Carn Brae extends to the Nine Maidens' stone circle. This line, from the hill summit to the Nine Maidens, forms the baseline for an isosceles triangle with England's southern point, Lizard Point, at the apex.

At the Nine Maidens stone circle, evidence of a method—in

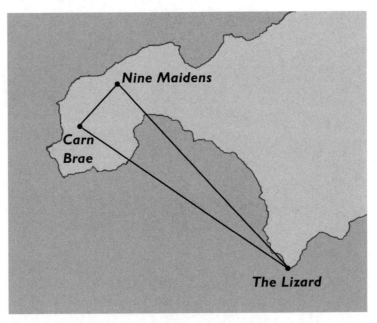

Figure AI.5. The line bearing 31.42 degrees joins Carn Brae summit with the Nine Maidens circle. This forms the baseline for an isosceles triangle with the tip of Lizard Point, the southern landmass extreme of Britain.

common with the other circles—is already indicated. On Google Earth the 31.42 bearing is exact, from the hill summit to the circle; all possible bearings to the circle fall within one-fifth of one degree. The legs of equal length are similarly accurate: one defines the landmass tip, and the other defines the limit of the rocks directly off the tip.* To conform with the geometry found at the Merry Maidens and the Boscawen-Un circles, there should also be a right triangle attached to the isosceles triangle.

The Nine Maidens stone circle is at right angles to Land's End and Lizard Point, but only when the skerries off both peninsulas are used in the geometry. It is actually possible to draw two right triangles from the Nine Maidens: the first has geometric points at (a) the base of the Land's End Peninsula, and (b) the skerries off Lizard Point; the second has geometric points on (a) the skerries off the Land's End peninsula, and (b) the tip of the Lizard Point peninsula. These two triangles are shown in figure A1.6, but they are too close together to distinguish on this scale; they both have a right-angle corner at the Nine Maidens circle.

The geometry created between topographical extremes at the Nine Maidens is, once again, a right triangle joined with an isosceles triangle, and again with one line bearing 31.42 degrees, the two triangles are shown in figure A1.7.

If it were true that any randomly chosen points would repeatedly produce joined isosceles and right triangles when drawing straight lines to topographical extreme points, then the series of geometric figures outlined above would illustrate nothing unusual. But when random points are investigated, such geometric figures are not common, nor are pi- and phi-digit measurements.

---

*At the southern limit of Lizard Point, a series of craggy rocks is barely separated from the mainland. These rocks span about one hundred meters from north to south. The northern and southern limits of these rocks are used as geometric points. The triangle is therefore not precisely isosceles; it requires a tip-to-base movement of the geometric point. The tip is the same distance from Carn Brae as the base is from the Nine Maidens. These types of measurement are common; they define tip and base points where the present day hilltops and headlands are found.

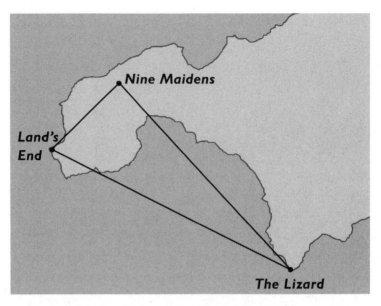

Figure A1.6. The right angles at the Nine Maidens joining England's western extreme at Land's End and England's southern extreme at Lizard Point.

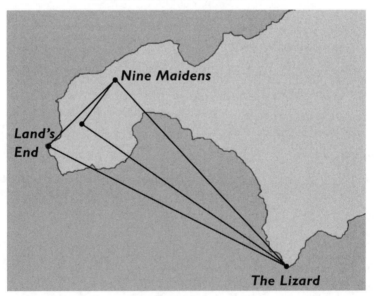

Figure A1.7. The short baseline of the isosceles triangle joins Carn Brae summit (the central unlabeled dot) with the Nine Maidens on a bearing of 31.42 degrees. The right triangle contains the isosceles triangle.

TREGESEAL EAST

There are two more circles in this western extreme of England. To the west of the Nine Maidens is the Tregeseal East stone circle. This circle is 6.00 nautical miles from the skerry off Hella Point and 4.5 nautical miles from the tip of Land's End.

A right triangle can be drawn between Land's End, the Tregeseal circle, and the eastern landmass extreme at Penlee Point.

In keeping with the geometry associated with the other circles, it should be possible to find the apex of an isosceles triangle at an extreme topographical point using one side of this right triangle as a baseline. This proves to be possible when a location on the northern peninsula of the landmass is adopted as a geometric point.

The joined right and isosceles triangles formed with the Tregeseal East stone circle are shown in figure A1.8.

The right triangle shown in figure A1.8 appears very precise, but while the isosceles triangle reaches an apex on the northern peninsula, it is not precisely at the tip or base, where the other points were found.

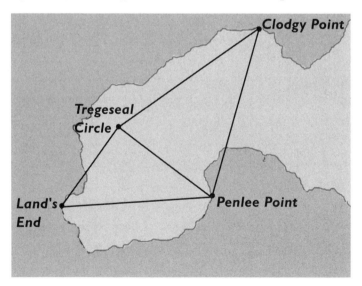

Figure A1.8. The eastern, western, and northern peninsulas of the landmass create joined right and isosceles triangles when the Tregeseal East stone circle is treated as a geometric point.

The difference amounts to nearly two hundred meters, and this lack of accuracy appeared uncharacteristic. Furthermore, there are no pi or phi digits associated with the geometric shapes shown in figure A1.8. There is however a much more precise northern apex point, and it is found more than one hundred kilometers distant on the eastern extreme tip of Lundy Island in the Bristol Channel.

The eastern peninsula on Lundy Island is equidistant between the Tregeseal circle and Penlee Point (the eastern landmass extreme shown in figure A1.8), thus completing the figure of joined triangles with near perfection. The small probability of locating an isolated yet adjacent landmass extreme point by this method is compounded by 31.42 degrees, which is the bearing from the Tregeseal circle to this extreme point on Lundy Island.

One leg of this elongated isosceles triangle therefore has a bearing of 31.42 degrees, and consequently, the same geometric motif and the same numerical motif appear at the Tregeseal stone circle just as they do at the Merry Maidens, Boscawen-Un, and the Nine Maidens (and at Stonehenge, Avebury, and numerous monuments described in the main text).

The apex of the elongated triangle on Lundy Island's distinctive eastern extreme is shown in figure A1.9.

The western line, taken from the stone circle, has a bearing of 31.42 degrees to an apex on Lundy's long peninsula. The point at the extreme tip of the peninsula is also found on a bearing of 31.42 degrees, but this time taken from the tip of the Land's End Peninsula. The same point on the peninsula tip on Lundy is found on a bearing of 16.18 degrees from the tip of the Lizard.

In each case the geometry associated with the stone circles opens a window on unusual measurements found between natural topographical extreme points. In this case the line bearing 31.42 degrees from England's western tip and the line bearing 16.18 from England's southern tip will intersect only once in the northern hemisphere, and this meeting point is precisely on Lundy Island's eastern tip. Following this lead, if a bearing of 31.42 degrees is taken from England's southern

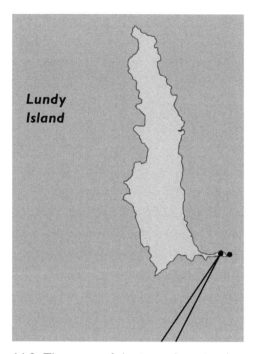

Figure A1.9. The apex of the isosceles triangle created on the fingerlike peninsula of Lundy Island. Penlee Point and the Tregeseal East circle are out of the image but are found at the two other corners of the triangle.

point at the Lizard, it leads directly to Brown Willy, Cornwall's highest point. And again if the same bearing is taken from Devon's southern extreme at Prawle Point, the line extends to Cleeve Hill summit, the highest point in the Cotswold Hills. (This line passes over Stanton Drew; see chapter 4.) The same bearing can be taken from a southern point on St. David's Head to Mount Snowdon and again from a southern coastal "point" on the Dingle Peninsula to Ben Nevis. From Ben Nevis a further line bearing 31.42 extends to a point on Cape Nordkinn at Europe's northern extreme. The point where this line crosses the northern coast is also found on a bearing of 31.42 from Scandinavia's highest summit, Galdhøpiggen. The monuments are apparently making reference to the strangely ordered nature of topographical extreme points as they are found today.

PORTHMEOR

The Porthmeor circle is found by extending the line from Carn Brae summit over the Nine Maidens. The distance between the two circles is 1.00 nautical miles, and the line has a bearing of 31.42 degrees from Carn Brae summit.

There are two skerries in Mount Bay to the south of Porthmeor. The first is reached with a bearing of 161.8 degrees; the other is 6.18 nautical miles from the circle. There is also a small island off the northern tip of Godrevy Head, and from this little lump of rock, it is 8.00 nautical miles to the Porthmeor circle, and it is also 8.00 nautical miles from this rock to the base of St. Michael's Mount, one of Cornwall's most famous landmarks.

St. Michael's Mount dominates the bay, a small mountain intermittently isolated from the landmass by tides covering a single causeway. A geometric point placed on this causeway, directly at the base of the mount, creates an isosceles triangle with the Godrevy rock island and

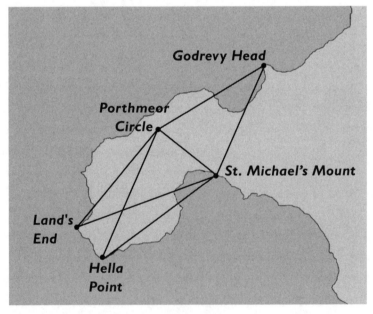

Figure A1.10. Geometry between extreme points and
Porthmeor circle: a right triangle and two isosceles triangles.

the Porthmeor circle. Hence, the two skerries in the bay are signified by phi-digit measurements, and the single large island in the bay is signified by geometry. The signified point is, in this case, at the base of the mountain on its northern limit. The significance of this point is found by joining it to the Porthmeor circle. This creates the baseline for a right triangle with the third corner at the tip of Land's End. The same line creates the baseline for an isosceles triangle with the apex at Hella Point, the southern landmass extreme.

Points on the southern and western limits of the landmass create an isosceles and a right triangle that share a line. The shared line extends from St. Michael's Mount to the stone circle at Porthmeor. This is also the baseline for an isosceles triangle with the Godrevy rocks 8.00 nautical miles to the northeast. In order to identify the natural geometry being described by the circle builders, the coastal points signified in this geometry are joined together independently.

It is 16.18 nautical miles from the tip of Godrevy Head to the Land's End Peninsula. Once again, by measuring between the natural points specified by the circle, pi and phi digits result. Move to the base of Godrevy Head, and it is 16.18 nautical miles to the tip of Hella Point. These results are echoed at each monument. They direct us to an investigation of extreme points where the subliminal geometry of the landmasses is spelled out. When measuring between these natural points in Cornwall, the result is strikingly geometric.

The highest point in figure A1.11, Carn Brae, and the eastern extreme at Penlee Point are both equidistant from the two western extremes of the landmass.

Two more geometric figures can be drawn, as shown in figure A1.12.

The location of the monuments is therefore cleverly duplicating the geometry created by joining the natural extreme points together.

Moving eastward in Cornwall fourteen stone circles on Bodmin Moor provide further clear evidence of this common code of practice at work. The highest point in Cornwall, Brown Willy, is located on Bodmin Moor.

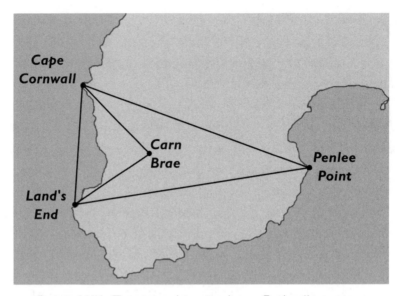

Figure A1.11. Two isosceles triangles at England's western extreme with the baseline joining Land's End and Cape Cornwall, and the apexes at Carn Brae and Penlee Point.

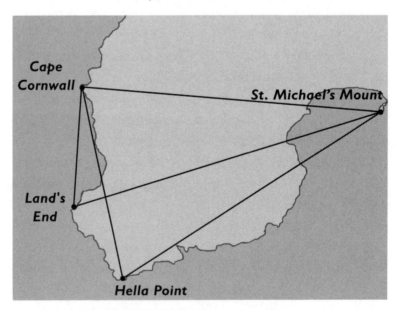

Figure A1.12. West Cornwall: The figure shows the two large triangles created by joining the four extreme points. One is a right triangle, the other an isosceles triangle. They share the line joining Cape Cornwall and St. Michael's Mount.

FOURTEEN STONE CIRCLES
ON BODMIN MOOR

Fernacre

The highest and largest among the ancient stone circles on Cornwall's Bodmin Moor is Fernacre, with a diameter of about 45 meters. Looking to the western coastline, some twenty kilometers from the circle, there are two notable peninsula points, Pentire Point and Trevose Head. These extreme points are highlighted in figure A1.13, and the location of the Fernacre circle is also shown.

The two peninsulas pointing west are highlighted in figure A1.13, and a similar southward-pointing peninsula on the south coast can also be seen, called Gibbon Head.

The western extreme point on Trevose Head is 16.18 nautical miles from the Fernacre stone circle.

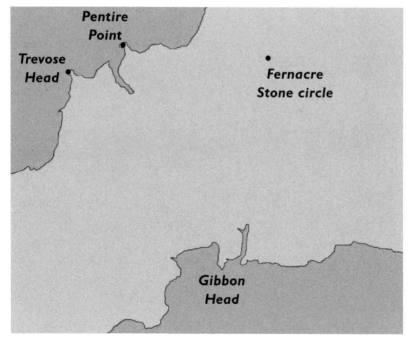

Figure A1.13. The Fernacre circle on Bodmin Moor with Pentire Point and, farther to the west, Trevose Head. A similar "pointy" headland, Gibbon Head, is labeled on the south coast.

The same distance can be measured from the Fernacre circle to the base of Gibbon Head in the south. From this base point the bearing to the Fernacre circle is 6.18 degrees.

Thus, a single point on the base of Gibbon Head is specified by these duplicate phi-digit measurements, and the line from this point to the tip of Trevose Head is the baseline for an isosceles triangle with the Fernacre circle at the apex. The triangle is shown in figure A1.14.

If the southern geometric point is moved from the base toward the tip of Gibbon Head, the isosceles triangle is broken, but the line from Gibbon Head then has a bearing of 314.2 degrees to the tip of Trevose Head.

The same geometric point on Gibbon Head is at the corner of a right triangle drawn between Trevose Head, Gibbon Head, and Britain's southern landmass extreme at the Lizard. Consequently, as

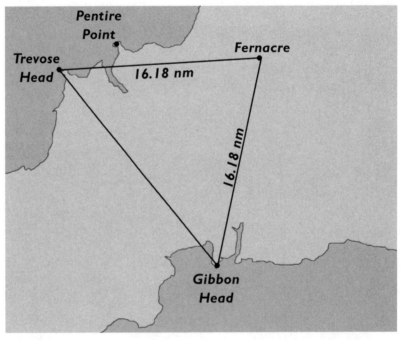

Figure A1.14. The isosceles triangle with the apex at the Fernacre stone circle and corners on the base of Gibbon Head and the tip of Trevose Head. The legs are 16.18 nautical miles long; the eastern leg has a bearing of 6.18 degrees to the stone circle.

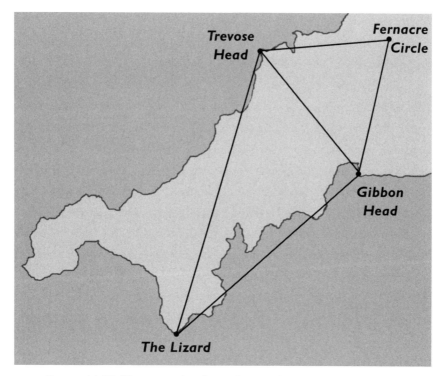

Figure A1.15. The joined right and isosceles triangles with the Fernacre circle at the isosceles apex.

with the circles already considered, by using geometric points on the peninsulas, a joined right triangle and isosceles triangle can be drawn between the three peninsulas and the stone circle; phi and pi digits result from measuring these lines. The two triangles are shown in figure A1.15.

Both of the geometric corners, and the line bearing 314.2 degrees, can be measured from points along the one-kilometer length of Gibbon Head.

The Fernacre circle is 0.75 nautical miles west of Bodmin Moor's highest summit, Brown Willy. Continuing westward on this line, from the western edge of the Fernacre circle, Google Earth gives a distance of one nautical mile (and four meters) to the eastern edge of the Stannon stone circle. The same distance separates the Porthmeor circle and the Nine Maidens.

Stannon

The Stannon stone circle is 16.18 nautical miles from the tip of the Bedruthan Peninsula, which is shown in figure A1.16.

The distance from the western tip of the Bedruthan Peninsula to the Stannon stone circle is 16.18 nautical miles (following the lower line in figure A16). The same distance is measured on the upper line from the western tip of Trevose Head to the Fernacre circle. The two circles are one nautical mile apart. A line 12.00 nautical miles in length can be drawn from the Fernacre circle, over the Stannon circle, to the tip of Pentire Point, the most northerly peninsula in figure A1.16. A joined right triangle and an isosceles triangle can be drawn between the Bedruthan Peninsula, Pentire Point, Gibbon Head, and the Stannon stone circle, as shown in figure A1.17.

In summary two isosceles triangles, both with legs measuring 16.18 nautical miles, join the Stannon and Fernacre circles with extreme points. In both cases another coastal extreme point creates a right

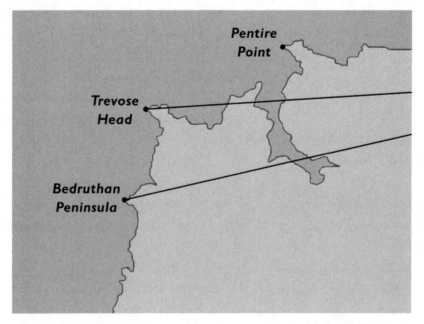

Figure A1.16. Two western extreme points are highlighted: Trevose Head and the Bedruthan Peninsula. The lines from these points to the Fernacre and Stannon circles both measure 16.18 nautical miles.

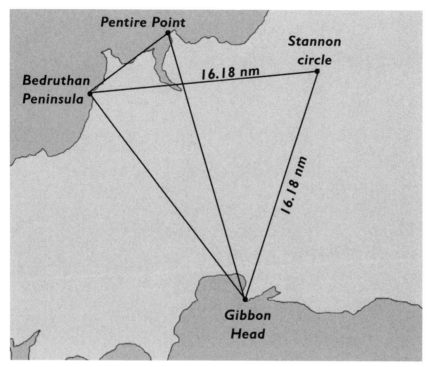

Figure A1.17. The isosceles and right triangles joining the Stannon stone circle, the Bedruthan Peninsula, Pentire Point, and Gibbon Head. The isosceles triangle's legs measure 16.18 nautical miles.

triangle attached to the isosceles triangle. In this way these two large circles on Bodmin Moor define geometry in the natural landscape.

East Moor and West Moor

Bodmin Moor covers roughly three hundred square kilometers, and its highest summit, Brown Willy, is on West Moor (High Moor). The highest summit on East Moor is found on a long, craggy outcrop called Kilmar Tor. To the south of Kilmar Tor is the most spectacular summit on Bodmin Moor, Stowe's Hill, where the massive flat boulders of the Cheesewring are stacked one upon another (see figure A1.18). These boulders (like Avebury's Sanctuary) align between England's eastern and western extremes.

The smaller circles on Bodmin Moor create the same geometric

Figure AI.18. The Cheesewring, Bodmin Moor.
Photo by Derek Winterburn.

motif as their larger neighbors, but this time the geometric points adopted are the highest points on the moor. The use of both high points and coastal extreme points as geometric points raises questions regarding the transient nature of coastal extremes.

Stripple Circle

The Stripple stone circle is due south of the Fernacre stone circle. The meridian joining Fernacre and Stripple is nearly five kilometers in length, yet the meridian alignment is perfect to within a few meters. This meridian line passes from the eastern limit of the Fernacre circle to the western limit of the Stripple circle henge. The extended line passes over two natural hills directly to the north and south of Fernacre; one is Rough Tor, the second highest summit on the moor, and the other is Garrow Tor. But this meridian line joining the circles does not pass exactly over the summit points of the two hills; it misses the summits by the diameter of the Stripple circle. Hence, these two natural summits appear perfectly aligned on a meridian that passes to the eastern limit of the Stripple stone circle. This line is 3.00 nautical miles in length. From this eastern point on the Stripple circle, a line bearing 90 degrees due east extends to the triangulation point on Kilmar Tor, the highest point on East Moor. The eastern limit of the Stripple circle therefore appears perfectly placed to create two right triangles aligned to the meridian, the first with Rough Tor and Kilmar Tor summits and the second with Garrow Tor and Kilmar Tor summits. On the Ordnance Survey 1:25,000 map of Bodmin Moor (map 109), these right triangles can be drawn by joining the point heights on the three hilltops and the eastern limit of the circle.

The highest summit directly to the north of Kilmar Tor is Hawk's Tor,* and this is at the apex of an isosceles triangle drawn with the summit of Garrow Tor and the Stripple circle. The sides of this triangle are 6.00 nautical miles in length, this time measured from the western extreme of the Stripple circle. Consequently, the meridian joining the Stripple circle and the Garrow Tor and Rough Tor summits forms the

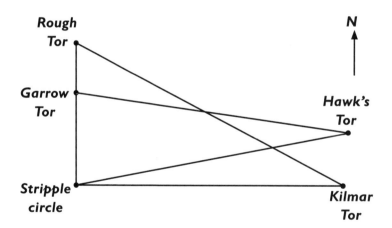

Figure AI.19. An isosceles triangle and a right triangle aligned to the meridian with the Stripple circle at the common corner. It is 3.00 nautical miles from the Stripple circle to the summit point on Rough Tor and 6.00 nautical miles from the Stripple circle to the summit point on Hawk's Tor.

---

*There are two summits called Hawk's Tor on Bodmin Moor, this one is the easternmost.

baseline for an isosceles triangle and a right triangle with the summits of Hawk's Tor and Kilmar Tor at the other two corners. All measurements are accurate using the Vincenty formula. Measurements using the Haversine formula and Google Earth differ from the Vincenty results by less than the circle's diameter.

The Stripple circle is one example of how the cardinal limits of the monument circle (or henge) define geometric points on the ground with considerable accuracy, and again the circle draws attention to the meridian alignment of two adjacent summits in the natural world.

Two Stone Circles: Leskernick Hill and Craddock Moor

Two stone circles on East Moor offer examples of how seemingly nondescript areas of ground can be recognized as unusual when they are seen in the context of topographical geometry. In both these cases the circle locations appear to have been decided for the same reasons.

From the first circle, on Leskernick Hill, it is possible to measure 1.618 nautical miles to the summit of Brown Willy. It is also possible to measure 3.142 nautical miles from this circle to the summit of the moor's highest northeastern point, marked "Ridge" on the Ordnance Survey map. The precise distance is marked by a large cairn on the ridge hilltop. Using only the diameter of the circle, a bearing of 16.18 degrees north of west can be taken from the Ridge to the circle, and a bearing of 6.18 degrees north of west from the circle to Brown Willy's summit.

The second circle is at Craddock Moor (part of East Moor); from this circle it is possible to measure 1.618 nautical miles to East Moor's highest summit, Kilmar Tor. A bearing of 3.14 degrees on this line reaches the center of the Kilmar Tor outcrop. It is also 0.618 nautical miles from the Craddock Moor circle to the nearest high point, the summit of Stowe's Hill.

In both cases there is only one small area of ground from which these measurements can be taken, and the two circles are located on each, as if to define the area of ground from which all these measurements can be accurately made. The additional significance of these small areas of ground is found by further measurements.

The nearest high point overlooking the Leskernick circle is Beacon Hill summit, 0.5 nautical miles away. The highest summit on the south of Bodmin Moor is Caradon Hill, and a line from this summit to the Leskernick circle has a bearing of 314.2 degrees. In both cases the high points are defined by precise measurement; they are then specified points. These points are joined, and the new line is measured in order to reveal topographical geometry. The distance between Caradon Hill summit and Beacon Hill summit is 6.18 nautical miles.

Once again phi digits are found by measuring between the natural topographical extremes specified by the monument. The area of ground defined by the Leskernick circle therefore becomes very unusual, even slightly magical.

The Craddock Moor Circle

This circle likewise draws attention to figures in the natural landscape. The process by which this is achieved is to specify points that appear geometrically arranged in the landscape. The first point specified with pi and phi digits is East Moor's highest point, Kilmar Tor.*

This craggy outcrop can be reached from the circle on Craddock Moor with a line 1.618 nautical miles long on a bearing of 3.142 degrees. The second specified point is Stowe's Hill summit (site of the Cheesewring), and 0.618 nautical miles can be measured from the circle to this summit.

Give or take a few meters, from the point specified on Kilmar Tor, the bearing to the summit of Stowe's Hill is 161.8 degrees. Therefore, the phi-digit bearing between these two natural points is indicated by the two phi-digit distances measured from the circle to them.

The natural geometry unfolds because the line bearing 161.8 degrees from Kilmar Tor to Stowe's Hill is about 1.3 nautical miles long, and the same distance can be measured from the Stowe's Hill summit to

---

*Measurements were taken from a point on the west side of the circle.

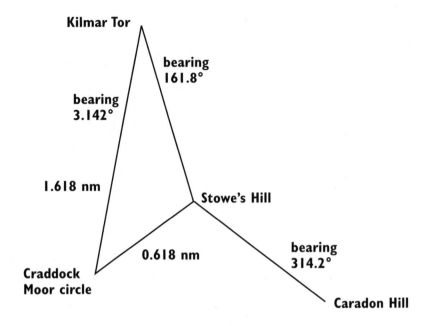

Figure A1.20. The Craddock Moor circle and
associated topographical geometry.

the summit area of Caradon Hill, the highest point on the south of the
moor. From this summit area it is possible to take a bearing of 314.2
degrees back to the summit of Stowe's Hill.

The end result is to discover that an isosceles triangle can be drawn
between three adjacent natural high points with Stowe's Hill at the
apex. One leg of the triangle has a bearing of 161.8 degrees to Stowe's
Hill summit, and the other has a bearing of 314.2 degrees to the same
point.

King Arthur's Down Double Circles

These two circles are side by side in a seemingly inauspicious location.
From the northern circle a line extends over the summit of Garrow
Tor, over the summit of Brown Willy, and continues to the rounded
summit of Buttern Hill, which is 3.12 nautical miles from the circle.
So the circle builders draw attention to the alignment of two of the
moor's highest summits with the highest summit of them all, Brown

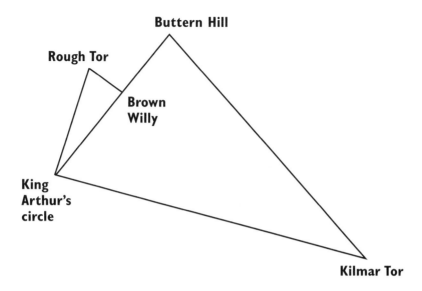

Figure A1.21. King Arthur's circle and associated
isosceles triangle and right triangle geometry.

Willy. But the wonder of this line is that it forms the baseline for a
right triangle with the highest summit on East Moor, Kilmar Tor, at
the other corner. The specific corner point on Kilmar Tor is at the
western base of the exposed rocky outcrop that constitutes the hill
summit.

The circles span about forty-five meters, and this small area proves
to be equidistant from the Brown Willy and Rough Tor summits, the
two highest points in Cornwall. Thus, the joined right triangle and
isosceles triangle are complete, with one line producing a pi-digit measurement. The key points in the geometry are the highest summits, as
shown in figure A1.21.

Goodaver

The Goodaver circle is the most centrally placed on Bodmin Moor.
It was once very dilapidated, and some stones may not have been re-
erected accurately. Beacon Hill is the high point to the north of the
circle. The bearing from this summit to the circle is 161.8 degrees.

The Goodaver circle is at right angles to summit points on Stowe's Hill and Brown Gelly, another local high point. The distance from the circle to these points is 3.00 nautical miles and 1.5 nautical miles, respectively. The two summits are therefore specified by this geometric arrangement. But why? When the two summits are joined west to east, the bearing between them is 90 degrees. Thus, the natural alignment of high points on the Earth grid is specified, because this line forms the hypotenuse of a right triangle with the Goodaver circle at the right-angle corner.

Extending this hypotenuse from Stowe's Hill over Brown Gelly leads directly to a summit point on Greenbarrow Downs. These three southern high points on Bodmin Moor are therefore aligned east to west. Once again, if the line from Stowe's Hill is extended, this time over the Goodaver circle, it leads directly to the summit of Garrow Tor. The joined right and isosceles triangles are now complete because summit points on Garrow Tor and Greenbarrow Downs have been specified by this process, and these two summit points prove to be equidistant from Stowe's Hill summit. The familiar signature to the work is discovered by measuring the lines. A distance of 3.142 nautical miles can be measured between the summit points on Greenbarrow Downs and Garrow Tor. The Goodaver circle therefore draws attention to the isosceles triangle created by these high points, with one leg aligned east to west and the baseline of this natural triangle measuring 3.142 nautical miles, as shown in figure A1.22.

The geometry in figure A1.22 appears to be all but perfect using the point heights marked on Ordnance Survey Explorer map 109 (Bodmin Moor 1:25,000). This is confirmed using computer-based measurements and Google Earth.

In summary the original Goodaver circle makers followed the code of practice in deciding this circle's location. The bearing from the local high point is 161.8 degrees. A further two high points are signified by measurement, one at 3.00 nautical miles distant, the other at half this distance. Joining these points with the Goodaver circle creates a right triangle with the hypotenuse aligned east to west. Extending the two

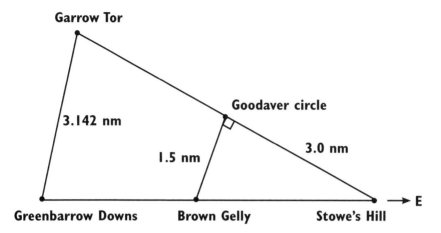

Figure A1.22. The Goodaver circle with an isosceles triangle and a right triangle, pi digits, and an east-to-west alignment.

long legs of the right triangle from Stowe's Hill summit finds two further summits, and these are 3.142 nautical miles apart and equidistant from Stowe's Hill. Thus, the Goodaver circle becomes integrated into the natural isosceles triangle of high points aligned east to west and in doing so creates a right triangle within the isosceles. This geometry between high points is almost identical to that found with other circles and coastal extreme points. If the coastal extremes are considered to be transient and this geometry was created by a conscious process, the builder's work presents the entire distribution of the landscape as something other than chaotic.

Leaze Circle

The nearest high point to Leaze circle is Garrow Tor, which is reached on a bearing of 31.42 degrees. The next nearest high point is the western Hawk's Tor, which is reached on a bearing of 161.8 degrees. A distance of 1.618 nautical miles can be measured between the two specified high points.

Louden Circle

A perfect square aligned to the meridian can be drawn with three corners on summit points located on Beacon Hill, Brown Gelly, and

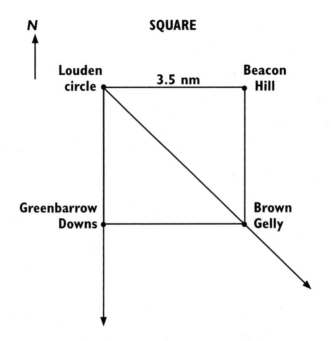

Figure A1.23. The square aligned to the meridian created
by three high points and the Louden stone circle.

Greenbarrow Downs, and with the fourth corner on the Louden circle. The sides of the square are 3.5 nautical miles long, and consequently, all three summit points can be defined with accuracy. Figure A1.23 shows a diagonal line running across the square, creating two right-angle isosceles triangles with the Louden circle at one corner.

Extending the diagonal line in figure A1.23 leads to a specific point just beneath the tip of Rame Head, a pronounced peninsula on the south coast. Extending the line due south of the circle and over Greenbarrow Downs leads to a peninsula point beside Gibbon Head at Polruan. A distance of 16.18 nautical miles can be measured between the two specified tip and base points on the coast.

Nine Stones

The small, restored Nine Stones circle to the northeast on Bodmin Moor yet again draws attention to natural landscape geometry, and again there is an isosceles triangle between high points. The circle is on the line joining the two nearest summits, the Ridge and Fox Tor. This line is 1.00 nautical mile long. The next high point in the chain is Hawk's Tor (east), and this summit, like Fox Tor, is also 1.00 nautical mile from the Ridge. The result is an isosceles triangle drawn between three adjacent natural high points with the two legs measuring 1.00 nautical mile each. The stone circle specifies two of these points by aligning between them.

With this circle it is hard to see any further justification for its location; a single alignment such as this could be coincidental. However, all the circles discussed so far identify geometry between topographical extremes measured in nautical miles.

Trippet, Westmoorgate, and the Hurlers

The Trippet stone circle is located about five kilometers south of Fernacre. It is hard to find any evidence that this circle was located with regard to topographical geometry. There also are two further circle sites on Bodmin Moor where the familiar right triangle and isosceles triangle geometry cannot be created nor are any alignments and familiar measurements apparent. The first is the Westmoorgate circle, where all the stones are recumbent. The second is the largest site on the moor, the three large adjacent circles that make up the Hurlers. The common hallmarks that define the great majority of stone circles of west Cornwall appear to be absent at the Hurlers.

Duloe

Finally, there is a small, isolated circle of stones at Duloe, about twenty kilometers south of Bodmin Moor and six kilometers from the south coast. Just off the coast is the small, isolated island called Looe (or St. George). Looe Island measures about 400 × 500 meters, a rough circle of land with a fingertip to the east and a chevron-shaped sand and shingle bank sitting like a party hat on its northern extreme.

A line from the western fingertip of Looe passing over the Duloe circle continues to the cairn on the summit of Brown Willy. The line is 16.44 nautical miles long, but if the distance is measured from the northern tip of Looe Island to the cairn on Brown Willy's summit, it is 16.18 nautical miles.

The regional high point to the northeast of the Duloe circle is Kit Hill; this summit can be reached with a bearing of 45.00 degrees. This line continues to Yes Tor, the second highest point in southern England. Extended to the south coast, this line reaches the peninsula of Pencarrow Head, which is roughly 300 meters square. From this lump of land, 31.42 nautical miles can be measured to the summit of Yes Tor; 16.18 nautical miles to the base of Kit Hill; and 6.18 nautical miles to the Duloe circle. Thus, the alignment of high points, coastal extreme points, and stone circles is played out on a small scale at Duloe using pi and phi digits.

APPENDIX 2

MAESHOWE DETAIL

The white line shown in figure A2.1 is the line of longitude, the meridian that runs directly through Maeshowe, also known as the Mound in the Meadow. The meridian passes over the southern extreme of the western half of Mainland at the Holm of Houton, indicated by the white dot in figure A2.1. The Holm of Houton is intermittently connected and disconnected from the mainland depending on the tide. It is a small and vulnerable area of land, and the meridian of Maeshowe cuts off the chevron-shaped peninsula of the island that points east. From the tip of the chevron, a second line has been drawn in black. This line also passes back to Maeshowe.

In figure A2.1 the two straight lines extend northward and cross over each other at Maeshowe; they then continue to Haafs Hellia, which is the northern limit of the landmass. The black line goes straight to the extreme tip of Haafs Hellia, and the white line passes to a base point, as shown in figure A2.2.

The black line reads from the south to the north: ^-^-^. The central symbol is Maeshowe, and the outer symbols are the landmass corners, both with chevron-shaped peninsulas indicating cardinal directions.

The line running from east to west through Maeshowe similarly identifies extreme points. The line bearing 90.00 degrees from Maeshowe is shown in black in figure A2.3.

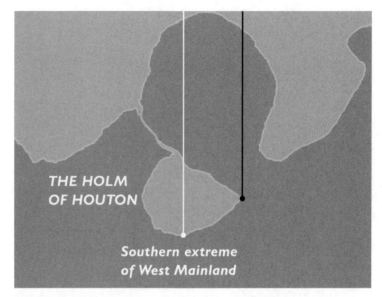

Figure A2.1. The Holm of Houton. The southern extreme
point is on the meridian due south of Maeshowe.

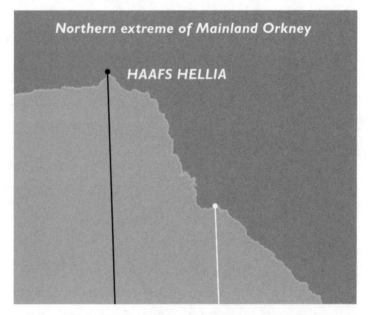

Figure A2.2. Haafs Hellia, the northern extreme of the
landmass of Mainland, is shown with the lines extended from
figure A2.1. The black line joins Maeshowe with the tips of
two chevron-shaped landmass extreme points.

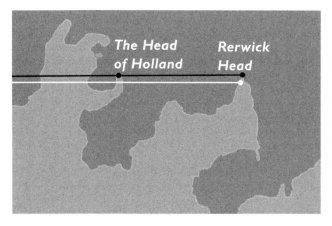

Figure A2.3. The Head of Holland and Rerwick Head, due east of Maeshowe.

The black line bearing due east from Maeshowe clips the tip of the strangely thumblike Head of Holland and just misses the tip of Rerwick Head. The white line is drawn from Rerwick Head back over Maeshowe. Figure A2.4 shows these lines, after they have crossed each other at Maeshowe, reaching the western landmass extreme of Mainland at Neban Point.

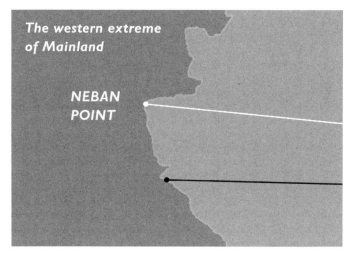

Figure A2.4. Neban Point on the western extreme of the landmass of Mainland. The black line is extended from the Head of Holland over Maeshowe to Neban Point, and the white line is extended from Rerwick Head also over Maeshowe to Neban Point. The lines identify the tip and a base point on this, the western extreme of the landmass.

THE CORNERS OF MAINLAND

In conclusion, the Mound in the Meadow is in the crosshairs of lines joining cardinal landmass extreme points, but in three cases the lines indicate both a tip and a base point on the landmass limit. It seems the people who located the monument, if they were not aware of these landmass points as they appear today, had nevertheless recognized a series of significant geometric points in the landscape. However, if there was no awareness, we must assume that the cardinal corners of the land-mass have eroded since Maeshowe was built and have by chance ended up due north, due south, and due west of the monument.

If the coastal extremes indicated by these alignments are considered as highlighted points specified by the monument, the specified points are then joined independently to reveal landscape geometry.

PI IN THE LANDSCAPE

Figure A2.5 shows four lines joining all four cardinal limits on the western Mainland landmass. There is a pi digit bearing on each line.

The box shown in figure A2.5 is created by two lines bearing 314.2 degrees and two lines bearing 31.4 degrees.* In three cases these lines join the tip and base coastal points specified by their cardinal alignment with Maeshowe. Figures A2.6a–d illustrate the extraordinary accuracy of the measurements; in these images the two highlighted points on each peninsula (found earlier by alignment with Maeshowe) are marked with dots. In each case, with near total accuracy, these specific points mark the beginning and end of the lines with pi-digit bearings.

At Haafs Hellia, the Holm of Houton, and Neban Point alike, the two dots defining the corners of the box are located (all but exactly) at the same points identified earlier by their alignments with Maeshowe.

---

*These lines were drawn using the Google Earth line facility; the bearings were then checked using the Haversine formula. This confirmed the accuracy on Google Earth's data to within a few meters on the ground.

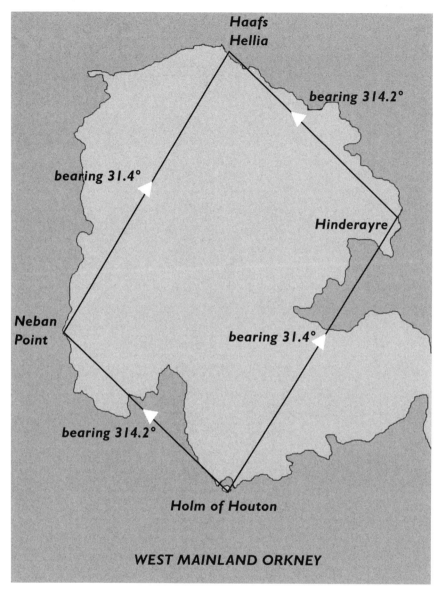

**Haafs
Hellia**

bearing 314.2°

bearing 31.4°

Hinderayre

**Neban
Point**

bearing 31.4°

bearing 314.2°

Holm of Houton

WEST MAINLAND ORKNEY

Figure A2.5. The four cardinal extreme peninsulas on the western
half of Mainland appear at the four corners of the box. Three
of these corners are specified by alignment with Maeshowe, the
fourth is at the eastern limit of this landmass just to the north of
Hinderayre Bay. Barely visible on the map, the slight extensions on
the corners of the box define, in three cases, the highlighted points,
shown earlier, on tips and flanks of the respective peninsulas.

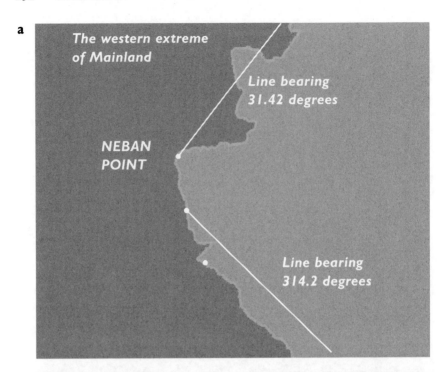

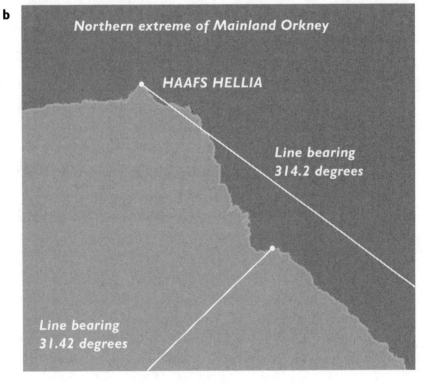

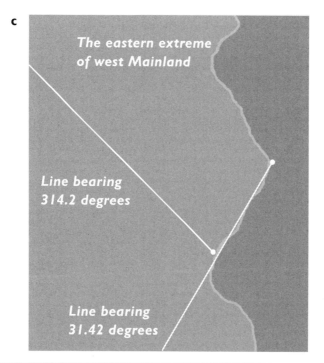

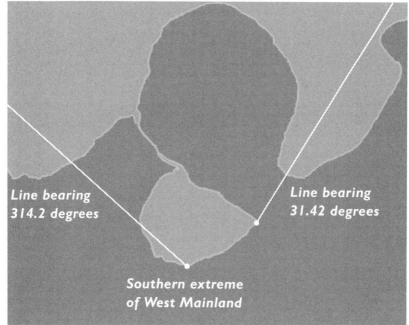

Figure A2.6 (a–d). The four cardinal extremes of the western half of Mainland, Orkney, in close-up.

This geometric figure was found by following the Neolithic method already recognized at numerous other sites. The same formula applies at almost every site: high points in the landscape and coastal extremes as we see them today are specified by alignment, and when these natural points are joined, they are found to be strangely ordered.

NOTES

FOREWORD.
KEYS TO EARTH'S SACRED GEOMETRY

1. See "What Is a Nautical Mile, and How Does It Differ from a Normal Mile and a Kilometer?" How stuff works, http://science.howstuffworks.com/innovation/science-questions/question79.htm (accessed August 30, 2015).

CHAPTER 1. SIGNPOSTS IN THE LANDSCAPE

1. Scott, *Hermetica,* 61.
2. Job 38:4, King James Bible.
3. Qur'an, 13:3, www.quran13/3.com (accessed August, 2015).
4. Rig Veda, book 5, hymn 85, translated by Ralph. T. H. Griffith, www.hinduwebsite.com/sacredscripts/rig_veda_book_5.asp (accessed July 14, 2015).
5. "The Book of Enoch. Chapters 61–105," The Reluctant Messenger.com, http://reluctant-messenger.com/1enoch61-105.htm#Chapter61 (accessed September 4, 2015), chapter 61, line 1.
6. Recinos, *Popul Vuh,* 169.
7. Bernardino de Sahagun, http://alternativearchaeology.jigsy.com/cult-of-the-plumed-serpent (accessed August 2015).
8. Roys, "Book of Chilam Balam of Chumayel," www.sacred-texts.com/nam/maya/cbc/ (accessed August 2015), 61.
9. Gaius Julius Caesar, *De Bello Gallico,* Book 6, Part XIV.
10. Stewart, *Nature's Numbers,* 5.

CHAPTER 2.
PATTERNS IN THE HILLS

1. Bayliss, Healy, and Whittle, *Gathering Time.*
2. "History of Silbury Hill," www.english-heritage.org.uk/visit/places/silbury -hill/history/ (accessed Aug 2015).
3. That ley lines do not exist was stated in Williamson and Bellamy, *Ley Lines in Question.* See also Johnson, *Archaeological Theory an Introduction,* 5.
4. Burl, *From Carnac to Callanish,* 155.

CHAPTER 3.
MEASURED MONUMENTS

1. Thom and Thom, *Megalithic Remains in Britain and Brittany,* 32.
2. Thom, *Chronicle.*
3. Burl, *From Carnac to Callanish,* 84.
4. Ibid., 70.

CHAPTER 5. CIRCULAR REASONING

1. Chris Veness, "Vincenty Solutions of Geodesics on the Ellipsoid," Moveable Type Scripts, www.movable-type.co.uk/scripts/latlong-vincenty.html (accessed June 29, 2015).

CHAPTER 6.
MOUNTAINS AND MONUMENTS

1. Chris Veness, "Vincenty Solutions of Geodesics on the Ellipsoid," Moveable Type Scripts, www.movable-type.co.uk/scripts/latlong-vincenty.html (accessed June 29, 2015).
2. Heath, *Sun, Moon and Stonehenge,* 76.

CHAPTER 7. ALMENDRES

1. Temple, *Egyptian Dawn,* 379.

CHAPTER 8. A LINE TO EVEREST

1. Burl, *From Carnac to Callanish,* 136.

2. Thom and Thom, *Megalithic Remains in Britain and Brittany,* 91.

3. Ibid., 65.

4. Boy and Allan, *Snowcaps on the Equator,* 31.

5. Michell, *New View over Atlantis,* 84.

6. Michell and Heath, *The Lost Science of Measuring the Earth,* 35.

7. Thom, "Megalithic Sites in Britain," www.spirasolaris.ca/sbb8a.pdf (accessed August 2015), 27.

CHAPTER 9.
INTERCONTINENTAL

1. Morton, Science in the Bible, www.christiananswers.net/ (accessed August 2015).

CHAPTER 11.
COPYING NATURE

1. Hancock, *Fingerprints of the Gods,* 170.

CHAPTER 12. COLD CONTINENT

1. Kuhn, *Structure of Scientific Revolutions,* 59.

2. Keay, *Great Arc,* 156.

3. "Mauna Kea," *National Geographic,* September 2012, supplement, Sperry and Vessels.

CHAPTER 13. ANGKOR

1. Heath, *Sacred Number,* Introduction.

CHAPTER 14. NEOLITHIC LANGUAGE

1. Graham Ellsbury, "The Turing Bombe," www.ellsbury.com/bombe1.htm (accessed July 14, 2015).

2. Smith, "Before Stonehenge," 26–48.

3. Santillana and Dechend, *Hamlet's Mill,* 6.

CHAPTER 15. PI AND THE SONGLINES

1. Chatwin, *Songlines,* 56.

CHAPTER 16. VIEWS FROM THE PAST

1. American Institute of Research, "Executive Summary."

2. Heath, *Alexander Thom,* 18.

3. Burl, *A Brief History of Stonehenge,* 267.

BIBLIOGRAPHY

American Institute of Research. "Executive Summary: An Evaluation of Remote Viewing: Research and Applications," September 29, 1995.

Bayliss, Alex, Frances Healy, and Alasdair Whittle. *Gathering Time: Dating the Early Neolithic Enclosures of Southern Britain and Ireland.* Oxford, England: Oxbow Books, 2011.

Boy, Gordon, and Iain Allan. *Snowcaps on the Equator: The Fabled Mountains of Kenya, Tanzania, Uganda and Zaire.* London: The Bodley Head, 1989.

Burl, Aubrey. *From Carnac to Callanish.* New Haven, Conn.: Yale University Press, 1993.

Charles, R. C. *The Book of Enoch.* London: Society for Promoting Christian Knowledge, 1917.

Chatwin, Bruce. *Songlines.* London: Vintage Classics, 2005.

Gaius Julius Caesar. *De Bello Gallico.* Book 6, part xiv.

Hancock, Graham. *Fingerprints of the Gods.* London, Heinemann, 1995.

Heath, Richard. *Sacred Number and the Origins of Civilization.* Rochester, Vt.: Inner Traditions, 2011.

Heath, Robin. *Alexander Thom: Cracking the Stone Age Code.* Pembrokeshire, Wales: Bluestone Press, 2007.

———. *The Sun, Moon, and Stonehenge.* Pembrokeshire, Wales: Bluestone Press, 1996.

Johnson, Matthew. *Archaeological Theory: An Introduction.* Oxford, England: Wiley-Blackwell, 2000.

Keay, John. *The Great Arc: The Dramatic Tale of How India Was Mapped and Everest Was Named.* London: HarperCollins, 2000.

Kuhn, Thomas S. *The Structure of Scientific Revolutions*. Chicago: University of Chicago Press, 2012.

Levi-Strauss, Claude. *The Savage Mind*. Chicago: University of Chicago Press, 1966.

Michell, John. *A New View over Atlantis*. London: Thames and Hudson, 2001.

Michell, John, and Robin Heath. *The Lost Science of Measuring the Earth*. Kempton, Ill.: Adventures Unlimited Press, 2006.

Morton, Joan Sloat. *Science in the Bible*. Chicago: Moody Press, 1978.

Recinos, Adrián, Delia Goetz, and Sylvanus G. Morley, trans. *Popul Vuh*. Norman: University of Oklahoma Press, 1991.

Roys, Ralph L. *The Book of Chilam Balam of Chumayel*. Washington, D.C.: Carnegie Institution, 1933.

Santillana, Georgio de, and Hertha von Dechend. *Hamlet's Mill*. Boston: Nonpareil, 1998.

Scott, Walter, trans. *Hermetica*. Great Britain: Solos Press, 1997.

Smith, Roff. "Before Stonehenge." *National Geographic,* June 2014, 26–48.

Stewart, Ian. *Nature's Numbers: The Unreal Reality of Mathematics*. New York: BasicBooks, 1995.

Temple, Robert. *Egyptian Dawn*. Great Britain: Arrow Books, 2011.

Thom, A., and A. S. Thom. *Megalithic Remains in Britain and Brittany*. Oxford, England: Oxford University Press, 1978.

Thom, Alexander. "Cracking the Stone Age Code." *Chronicle*. BBC-TV, October 1970.

———. *Megalithic Sites in Britain*. Oxford, England: Oxford University Press, 1967.

Vessels, Jane. "Mauna Kea," *National Geographic,* September 2012.

Williamson, Tom, and Liz Bellamy. *Ley Lines in Question*. West Sussex, U.K.: Littlehampton Book Services Ltd., 1983.

Wilson, Colin, and Rand Flem-Ath. *The Atlantis Blueprint: Unlocking the Ancient Mysteries of a Long-Lost Civilization*. Surrey, U.K.: Delta, 2002.

INDEX

Numbers in *italics* indicate illustrations.

BOOKS OF RELATED INTEREST

How the World Is Made
The Story of Creation according to Sacred Geometry
by John Michell with Allan Brown

The Dimensions of Paradise
Sacred Geometry, Ancient Science, and the Heavenly Order on Earth
by John Michell

Atlantis beneath the Ice
The Fate of the Lost Continent
by Rand Flem-Ath and Rose Flem-Ath

Göbekli Tepe: Genesis of the Gods
The Temple of the Watchers and the Discovery of Eden
by Andrew Collins

Secrets of Ancient America
Archaeoastronomy and the Legacy of the Phoenicians,
Celts, and Other Forgotten Explorers
by Carl Lehrburger

Point of Origin
Gobekli Tepe and the Spiritual Matrix for the World's Cosmologies
by Laird Scranton

Forgotten Civilization
The Role of Solar Outbursts in Our Past and Future
by Robert M. Schoch, Ph.D.

The Soul of Ancient Egypt
Restoring the Spiritual Engine of the World
by Robert Bauval and Ahmed Osman

INNER TRADITIONS • BEAR & COMPANY
P.O. Box 388
Rochester, VT 05767
1-800-246-8648
www.InnerTraditions.com

Or contact your local bookseller